THE
PAPERMAKER'S
COMPANION

The Ultimate Guide to Making and Using Handmade Paper

Helen Hiebert

Author of *Papermaking with Plants*

STOREY BOOKS

Schoolhouse Road
Pownal, Vermont 05261

The mission of Storey Communications is to serve our customers
by publishing practical information that encourages personal independence
in harmony with the environment.

Edited by Deborah Balmuth, Eileen M. Clawson,
  and Maryann Teale Snell
Cover design by Meredith Maker
Cover illustration by Laura Tedeschi
Text design by Susan Bernier
Production assistance by Jennifer Jepson Smith
Illustrations by Alison Kolesar
Indexed by Nan Badgett / Word•a•bil•i•ty

The information in this book is true and complete to the best of our knowledge. All recommendations are made without guarantee on the part of the author or Storey Books. The author and publisher disclaim any liability in connection with the use of this information. For additional information please contact Storey Books, Schoolhouse Road, Pownal, Vermont 05261.

Storey books are available for special premium and promotional uses and for customized editions. For further information, please call Storey's Custom Publishing Department at 1-800-793-9396.

Printed in the United States by R.R. Donnelley

10 9 8 7 6 5 4 3 2 1

Library of Congress Cataloging-in-Publication Data

Hiebert, Helen, 1965–
    The papermaker's companion: the ultimate guide to making and using handmade paper / Helen Hiebert.
        p.   cm.
    Includes bibliographical references and index.
    ISBN 1-58017-200-8 (alk. paper)
    1. Paper, Handmade. I. Title.
TS1124.5 H53 2000
676'.22—dc21   99-087351
                                                                    CIP

DEDICATION
▼▼▼

To Willam Jakob:
May your life be full of moments
of awe and discovery, and may you inspire us
big people to look at the world in new ways.

ACKNOWLEDGMENTS
▼▼▼

I am grateful to the many people who made this book possible:

To papermakers past and present who are keeping this craft alive and thriving by developing innovative ways to work with paper: thank you for sharing your ideas with me and nurturing me — you have helped shape my career.

To Pat Almonrode, Eugenie Barron, Andrea Peterson, and Paul Wong, who reviewed my manuscript; to Joyce Schutter, who shared her entire thesis on pulp spraying with me; to Mary Michaud, who helped me prepare the projects for the illustrator; and to the many papermakers who contributed their tips, tricks, and recipes: you are credited within the pages of this book.

To my husband, Ted, for his endless proofreading patience; and my new baby, Willam, who pushed me to make my deadline, because he was due two weeks before my manuscript!

To the New York and Portland Re-evaluation Counseling communities: you have quietly encouraged me to reach for my dreams, and your support has kept me going.

To Deborah Balmuth, my editor at Storey Communications: you are a delight to work with.

And to my students, readers, and future papermakers: your enthusiasm inspires me. I am privileged to work with you and learn from you.

Contents

INTRODUCTION TO PAPERMAKING

Paper is something we touch so many times a day — whether reading the newspaper, pouring cereal from a box, writing a letter, or sending a fax. We couldn't live without it; yet how many of us even think about how it's made? Modern, machine-made papers are produced using the same principle developed almost 2,000 years ago. In fact, a surprising number of people still practice this age-old craft.

THE HISTORY OF PAPER

The research and writings of Dard Hunter, a craftsman, papermaker, and scholar who traveled and made paper in the first half of this century, provide valuable reference material on the subject of papermaking. Hunter documented his findings in his book *The History and Technique of an Ancient Craft* (see reading list). Paper historians have continued to research the history of the craft, and other accounts of papermaking's history can be found; but no single work is as comprehensive as Hunter's.

Documenting World Traditions

The Research Institute of Paper History, in Brookline, Massachusetts, is run by Elaine Koretsky, who is something of a present-day Dard Hunter. Elaine has traveled all over the world since 1976, especially in Asia, where she has found an unbroken tradition of papermaking in countless remote villages. In particular, she has

WHAT IS PAPER?
True paper is made from a raw material that has been macerated (beaten) and broken down into tiny fibers, mixed with water, and formed into sheets on a screen surface that catches the fibers as the water drains through it. The individual fibers interlock and form a sheet of paper when pressed and dried.

concentrated on China. She documents the vestiges of old paper-making that still exist, collecting video footage, tools, and papers from villages where paper has been made for hundreds of years.

During visits to Burma over the years, Koretsky has collected all the tools necessary to produce the extraordinary bamboo paper that is used as a substrate in the beating of gold leaf. With her daughter Donna she wrote *The Gold Beaters of Mandalay* (see reading list), which describes this amazing process. One of the most interesting books in her collection is a Cambodian prayer book, probably from the 14th century. An accordion-folded handmade paper book, it has magnificent calligraphy and paintings in color on many of the leaves. She discovered a "twin" to this book at the Smithsonian Institution several years ago.

The Research Institute of Paper History includes a library of old, modern, and rare books dealing with paper history, the technology of papermaking, and manuscripts illustrating book forms in many ancient cultures. It also includes a collection of handmade papers, tools, equipment, and artifacts relating to historical papermaking, which have been gathered from all over the world.

Early Writing Materials

Many substances were used as writing surfaces before paper was developed. As early as the 14th century B.C., the Chinese inscribed marks on bones. In Egypt, hieroglyphs were carved in monuments of stone and written on papyrus. Ancient Babylonians carved their cuneiform characters into clay bricks and tablets using bone tools. Several ancient civilizations recorded information on metals such as brass, copper, bronze, and lead. Long before the time of Homer, written information was scratched into pieces of wood that were covered with wax, chalk, or plaster. In early Rome and countries in the Near East, palm and other leaves were etched with a metal stylus, and the scratch marks were filled with a carbon paint.

Our ancestors used many different writing surfaces, including clay tablets, parchment scrolls, and even birch bark.

Many cultures used the bark of trees as a writing surface: the ancient Latins used inner bark; the American Indians wrote on hammered bark of white birch trees; the aborigines of Central America, South America, and Mexico beat the inner bark of moraceous trees (from the mulberry family). Parchment, which became a common writing and printing surface in Europe, was made from sheep or goat skin. Parchment was probably used as early as 1500 B.C. in Pergamum in Asia Minor, but was not considered a common writing surface until 200 B.C. Vellum, another writing surface, was made from the unsplit skin of young calf, goat, and lamb.

Papyrus and Bark Papers

The word for papyrus in several languages *(paper, papier, papel)* comes from the Greek word *papyros.* But papyrus is not a true paper. Although the papyrus plant can be macerated and made into sheets of paper, traditional papyrus was made by cutting strips of the plant's stalk into lengths and overlapping them side by side. A second layer of overlapping strips was placed on top of and perpendicular to the bottom layer. These two layers were then pounded and laminated together.

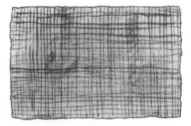

Traditional papyrus is not considered a true paper.

Substances similar to papyrus were made by the Mayans and the Aztecs, and the tradition may be as old as the aboriginal civilizations in Mexico. These substrates were similar to papyrus in that they were made by overlapping pieces of the inner bark of trees and pounding them until they were flat, but they were made from different plants. The Mayan substrate, called *huun,* was made from the bark of a ficus tree, and the Aztecs' *amatl* was crafted from mulberry and wild fig tree barks. In the Pacific Islands, bark papers called *tapa* were also made, and the people there decorated them and used them for clothing and for many domestic purposes. Some of these bark papers are still produced today.

The construction of bark papers is similar to that of papyrus.

WESTWARD MIGRATION

Ts'ai Lun, an emperor's eunuch in China, has been credited with paper's invention in the year A.D. 105, but there is some dispute over the exact date of origin. Earlier paperlike fragments made from hemp have been found in China, leading some historians to conclude that paper was invented at least two centuries earlier, but Ts'ai Lun most likely refined the process and promoted paper as a writing material.

By A.D. 615 the process had spread along the silk and trade routes to Japan. In about A.D. 750, it is believed that some Chinese papermakers were captured when the Turks defeated Samarkand and were forced to divulge their techniques, which had been kept secret. From there papermaking spread throughout the Arab regions, then to Egypt in the tenth century, and on to Morocco. Papermaking is believed to have been introduced in Europe first in Xativa, Spain, in 1151. England's first paper mill was established in about 1488, and papermaking was introduced to the United States by German immigrant William Rittenhouse in 1690 in Germantown, Pennsylvania.

PAPERMAKING MATERIALS

Before the invention of paper, writing in China was done on bamboo strips or pieces of silk. Bamboo was heavy and awkward, and silk was expensive; so alternative writing surfaces were sought. Ts'ai Lun's paper was probably made from a mixture of tree bark, discarded cloth, fishnets, and hemp waste.

Native plants were used to make paper in many countries. In China and Japan, mulberry was a common fiber that made a thin, soft, absorbent paper suitable for calligraphy and woodblock printing. Hemp and flax were used in Samarkand because they were in abundant supply. In Europe, papermakers began making pulp from old rags, an abundant raw material that took less time to reduce to a pulp than raw hemp, flax, or cotton.

Experimenting with Plants

As printing technology developed and books and newspapers came to be in demand, it became difficult to find enough rags for papermaking. The use of the wood from trees for papermaking was first suggested in 1719 by René de Réamur, an entomologist who observed that wasps made their nests out of regurgitated wood. Noting the paperlike quality of wasp nests, he surmised that paper could be made using a similar technique on a much larger scale.

Réamur never tested his theory, but his investigations did prompt others to try to make pulp from materials other than rags. Several people experimented with vegetation, including conferva (swamp moss), broom, moss, cabbage stalks, potatoes, cattails, and many other plants. In Germany, Jacob Christian Schäffer (1718–90) did extensive papermaking experiments using various plants from his own garden and neighboring fields. While Schäffer attempted to bring attention to the vast variety of vegetation that was available for papermaking, most of his samples contained 20 percent cotton rags, which he believed was necessary to bind the plant fibers together.

According to papermaker Dard Hunter, a Frenchman named Léorier Delisle produced the first pure plant paper in the West for a small book edition in 1786. He used the bark of the lime tree for the paper in one copy and the marshmallow plant in a second copy. Still others experimented with making paper from various plants, until finally in the early 1800s, Englishman Matthias Koops made use of plant fibers on a commercial scale using straw.

PAPER MADE BY MACHINE

All paper was made by hand until 1798, when Nicholas Louis Robert invented a papermaking machine in France. The machine was perfected in England in the early 1800s by the Fourdrinier brothers; to this day, paper machines bear the name Fourdrinier. On this

Wasps' nests inspired the production of paper from wood.

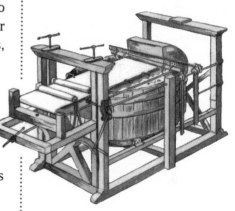

The first papermaking machine was refined in England in the early 19th century.

Calender

to smooth a paper's surface by passing the sheets between metal rollers.

machine, paper is formed on an endless wire screen, which travels over rollers and is then transferred onto a continuous felt, which then moves through a series of rollers that press, dry, **calender** (finish the surface), and wind the paper into a roll.

TWENTIETH-CENTURY REVIVAL OF HAND TECHNIQUES

In the first part of the century, Dard Hunter revived the craft of hand-papermaking and traveled extensively around the world, studying historical and contemporary papermaking practices and building a comprehensive collection of papermaking tools and paper samples. His collections and archives are currently housed at the Robert C. Williams American Museum of Papermaking in Atlanta, Georgia.

During the hand-papermaking renaissance of the 1960s and 1970s, artists began painting and sculpting with paper pulp, and making custom sheets for their prints. By offering apprenticeship and internship programs, small hand paper mills pass on this knowledge and help preserve an age-old tradition.

Entrepreneurs are using handmade paper to develop stationery, books, lamps, jewelry, and unique handmade sheets. Innovative programs are taking place in developing countries — both to ensure that the tradition of hand-papermaking is preserved and to create job opportunities.

EDUCATIONAL OPPORTUNITIES

Several universities and colleges now offer papermaking, and classes and workshops can be found at many community centers and private studios. *Hand Papermaking*, a quarterly journal, has listings of resources around the country.

CONTINUING THE TRADITION

In this book, I hope to share many tips and tricks I've learned over the years. You, too, can create unique textures, colors, and shapes in paper and then manipulate the medium to create books and three-dimensional works. The possibilities are amazing — and the papermaking renaissance is still young — so grab your mould, charge your vat, and join in!

Papermaking
Basics

PART 1

CHAPTER 1

GETTING EQUIPPED

The basic papermaking process involves dipping a screen stretched across a frame (the mould and deckle) into a vat of pulp, lifting the screen out of the vat, and shaking it back and forth — and side to side — so that the fibers interlock and bond on top of the screen surface as the water drains through the screen. The freshly made sheet of paper is then **couched** (transferred) onto a surface — usually a **felt** — and is then pressed and dried.

You can start making paper with simple equipment, which you can purchase or build yourself. As you become more serious about the craft, there are specialized items you might wish to acquire. I prefer to start small and build up as necessary. Following is a description of the basic equipment you will need to get started, as well as suggestions for upgrading to more sophisticated equipment.

People in this field are always inventing new equipment, adapting items from other industries, and redesigning old machines to function better. Networking is a great way to find the things you need or learn how to build them. There are several papermaking organizations, a trade magazine called *Hand Papermaking,* and a wealth of information to be found on the Internet, from papermaking Web sites that highlight artists' offerings to those that show the uses of the process in the developing world to sites that feature university and art center course offerings.

Couching

(pronounced cooching) transferring a freshly made sheet of paper from the mould surface onto a felt, a sheet of nonwoven polyester interfacing, or another surface.

Felt

a woven wool fabric with a smooth, brushed surface and absorbent quality that aids in releasing a sheet of paper from the mould.

PAPERMAKING PROCESSES

There are many variables in the papermaking process, and there is no single way to make a sheet of paper. From collecting the fiber to drying the finished sheets, you can alter the way in which you process your materials.

Fiber Collection Options
+ Forage
+ Purchase from store or papermaking supplier
+ Harvest plants
+ Gather recycled materials

Precook Preparation Options
+ Steam, strip, and scrape bast fiber
+ Ret or decorticate tough leaf fiber
+ Presoak all plant fiber

Cooking Options
+ Cook plant fiber in soda ash, wood ash, washing soda, or lye
+ Cook rags in soda ash to get rid of additives or to speed up beating process

Beating Options
+ Hand-beat with mallets
+ Stamp
+ Blend
+ Process with a paint mixer, Whiz Mixer, hydropulper, or Hollander beater

Sheet Forming Options
+ Western
+ Eastern
+ Deckle box
+ Variations

Pressing Options
+ Sponge, rolling pin, or brayer
+ C-clamps and plywood boards
+ Recycled press (flower press, cider press, book press, etc.)
+ Hydraulic press

Drying Options
+ Air dry
+ Board or wall dry
+ Exchange dry
+ Restraint dry

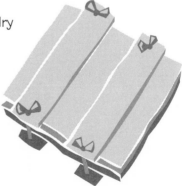

Fiber

cells that impart elasticity,
flexibility, and tensile strength
to plants.

ARRANGING YOUR SPACE

Since papermaking involves several steps, you will want to design
your work space to provide a smooth flow from one step to the next.
You will need space for storing dry papers and **fibers,** a place to
cook if you use plant fibers, and a space for beating the pulp, then
making, pressing, and drying the sheets. Each of these processes
requires various pieces of equipment and space. Here are some
things to consider when arranging your space:

WET AND DRY WORK AREAS

You will need a dry space in which to store fiber and tools when they
are not in use. You will also need a wet area for beating pulp and for
forming and pressing your paper sheets.

Your fiber needs to be kept dry so that it will not mold. Papermaking equipment, especially those pieces used regularly, such as
moulds, deckles, and felts, should be dried out between uses. Your

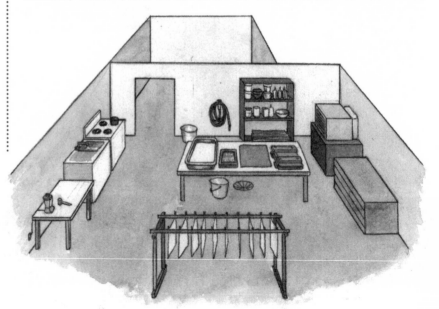

equipment can be stored off the ground in a wet work area. A lot of water is involved in preparing pulp and making papers, so working on a surface that can get wet is convenient. Drains in the floor are even better. You might also consider working outside (weather permitting, of course!).

VENTILATION AND ELECTRICITY

If you use plant materials, you'll need a stove or hot plate for cooking fiber. Fumes are released from cooking with **alkalis,** and it's best to vent them out of your work area. A hooded stove will do the trick. You can also work outside on a hot plate or a camping stove. If you work with dry pigments, you should wear a mask and work in a well-ventilated area.

If your cooking area is in your wet area, protect any electrical outlets from water. If possible, have the outlets at waist level rather than close to the floor. You may need electricity for hot plates, blenders, and other equipment.

WATER SOURCE

A faucet is the bare minimum you'll need for your water source, and a large sink can be handy. If you have a garden hose with a spray nozzle, you can take the water where you need it and use it when you want to.

NOISE

The beating process, in particular, can be noisy! If you prefer to pound your fiber by hand, or use a blender or other pulp reconstitutor, the noise will be a mild nuisance — not much louder than a washing machine.

But if you use a Hollander beater to prepare the pulp, you may prefer to put it in the garage or basement, or build soundproof walls. I recommend wearing earplugs (or protective headsets) when you are working close to the beater.

Alkali

a base substance that aids in the removal of noncellulose materials from a plant fiber when soaked or cooked in solution with the fiber.

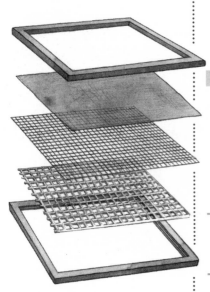

Before you build your mould and deckle, consider the many options available.

MATERIAL OPTIONS

When making your own mould and deckle, there are many things to think about — including size, wood, joints, screen, and support. Here are some considerations in selecting materials.

Material	Options
Size and type of wood	• Determine the size of paper you wish to make. When buying or cutting your wood frame, make sure you make the pieces large enough to account for the joints. • Use nonwarping, waterproof wood, such as mahogany (or seal the wood with polyurethane, Thompson's Water Seal, linseed oil, or tung oil).
Corner joints	• Prejointed canvas or needlepoint stretchers • Brass screws (rustproof) • Half-lap, dovetail, or other wood joints
Screen surface options	• Aluminum window screen • Brass or copper mesh screen (30 to 40 wires per inch) • Simple removable su and sha, made with a bamboo placemat or interfacing and no-see-um netting • Heat-shrinking polypropylene screening (available from papermaking suppliers)
Screen support	• Hardware cloth • Wooden ribs (for larger moulds) • Plastic grid, used in fluorescent lighting fixtures (for larger moulds)
Fasteners for attaching the screen to the wood	• Brass shim or aluminum flashing (used for chimneys) and escutcheon pins • Waterproof staples (Monel brand) and duct tape • Epoxy (for use with heat-shrinking polypropylene screening)

THE MOULD AND DECKLE

The *mould* is a screen surface stretched across some sort of frame. The *deckle* is a second frame without a screen that sits on top of (and is sometimes jointed to fit over) the mould to contain the pulp when it is dipped into the vat. The deckle controls the shape, size, and thickness of the sheet of paper formed on the mould.

There are three basic types of moulds and deckles. The Western mould can range from window screening stretched over canvas stretcher bars, to hardwood moulds and deckles with a sophisticated ribbing structure designed to hold the screen in place. The Eastern mould, the most common being the *sugeta* of Japan, differs from the Western version in that it consists of a flexible screen that is sandwiched and held taut in a hinged frame. The deckle box, originally from Asia, is a deep mould and deckle into which pulp is poured.

The mould acts like a sieve as paper is formed.

Western Mould and Deckle

There are two types of traditional Western moulds, which vary in screen surface: laid and wove. On a traditional laid mould, a series of parallel wires (laid lines) creates the mould surface. These wires are sewn into a mesh and then sewn onto wooden ribs, spaced at approximately 1-inch intervals, that support the wires from below and prevent them from sagging under the weight of the pulp and water. The lines created where they are sewn to the ribs are called the chain lines. These laid and chain lines are visible in sheets of dry paper that are formed on the mould.

The Western laid mould and deckle was developed in Europe in the 12th century, when papers were first made from old rags. The wove mould, which has a woven mesh surface like a fine window screen stretched across its frame, was developed later, in the 18th century. The surface of a sheet made on a wove mould is less distinctive and smoother than the laid surface.

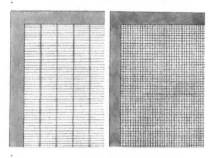

The laid surface (left) creates laid and chain lines, while the wove surface (right) makes paper with a consistent surface.

Make Your Own Western Mould and Deckle

You can make a simple and inexpensive mould and deckle using canvas stretcher bars from an art supply store. They come in many sizes. I recommend making this type of mould and deckle no larger than 8½" x 11".

If you decide to make a mould and deckle larger than 8½" x 11", you will need to support the underside with ribs or plastic grid. This may require making a rabbet groove for the plastic grid to sit on or cutting slots in the wood for the ribs.

Use a plastic grid to support the mould and deckle if you make one that's bigger than 8½" x 11".

You Will Need:

4   canvas stretcher bars measuring 14" x 1½"
4   canvas stretcher bars measuring 11½" x 1½"
Water- and weather-resistant wood glue (available at
      hardware stores)
Water-based polyurethane (oil-based is okay, but I don't
      like the fumes)
Small foam or bristle brush
Tin snips for cutting wire
Rustproof wire screen* (window screen or finer), cut to
      13" x 10½"
Hardware cloth (a heavy wire mesh with ¼" mesh squares)
      cut to 13" x 10½"
Staple gun and rustproof staples (Monel brand)
Duct tape
Scissors
¼" or ½" foam insulating strips (weather stripping)

There are alternatives to wire screening. Most papermaking suppliers carry a heat-shrinking polypropylene screening (and a special epoxy paste) that can be glued to the frame and tightened with the heat of a blow dryer.

1. Apply wood glue to the joints, and fit them together to form two wooden frames. Make sure the corners are square.

2. Apply polyurethane with a brush to waterproof the frames, and let them dry overnight. I recommend applying a second coat after the first has dried.

3. To make the mould, place the window screen on top of the hardware cloth, and lay the combined screen onto one of the frames. Center it and staple it to the frame at 1" intervals, keeping it taut. Put duct tape over the stapled edges to seal the screen and prevent pulp from slipping between them. This also protects your hands from the sharp edges of the screens.

4. To make the deckle, cut the foam strips to fit along the inside edge of the "window" on the other frame. Use a staple gun to attach the foam strips at 1" intervals. Since most canvas stretchers have a slight lip on the outer edge, these foam strips prevent pulp from slipping underneath the deckle onto the unscreened area of the mould.

step 1

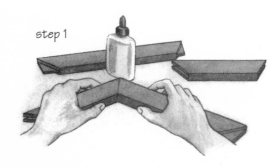

step 3

step 4

Eastern Sugeta

Sugeta is the Japanese word for the traditional papermaking tool in Japan. The *su* is the screen, which differs from the Western mould in that it is flexible. The *keta* is a hinged frame that holds the *su*. The *sugeta* is deeper than the Western deckle, to accommodate the different sheet-forming process (see Japanese papermaking, page 98). This process is more suitable for some plant fibers than the Western process, and the flexibility of the Japanese screen makes couching easier with many fibers, especially when you are forming thin sheets.

The traditional Japanese *su* is made from bamboo splints woven together with silk or nylon threads. Its design was based on earlier moulds made in China and Korea. Some *sus* are lined on one side with a **sha,** or fine silk mesh.

It is hard to find traditional *sugeta*s in this country, but a few crafts people are making contemporary versions (see resource guide).

Sha

traditionally in Japan, a silk layer sewn to the bamboo *su* to create the surface on which the sheets of paper are formed.

SIMPLE MOULD AND DECKLE

If you don't want to go to the trouble of making or purchasing your own mould and deckle, start simply. Make a simple screen with no deckle by cutting a piece of hardware cloth and a piece of window screening to the same size. Tape all the edges together with duct tape. You can even make shaped moulds this way (see chapter 10, Making Shaped Sheets, and chapter 15, Children's Projects, for other shape ideas). This is a deckle-less mould. The purpose of the deckle is to create a nice edge, as well as to control the shape and thickness. If you don't use a deckle, you will get a more feathered edge, and the thickness of your sheet will be harder to control, but nevertheless, you will be able to make paper.

MAKE YOUR OWN EASTERN SUGETA

You can make a simplified *sugeta* that is similar to the traditional Japanese *sugeta* from common materials such as bamboo placemats or sushi rolls and needlepoint stretcher bars.

You Will Need:

4 lengths of pine measuring 12" x ¾" x ¾" (prejointed needlepoint stretcher bars, available at sewing and craft stores)

4 lengths of pine measuring 16" x ¾" x ¾"

Water- and weather-resistant wood glue (available at hardware stores)

8 1¼" brass or stainless steel screws

Drill and drill bit (smaller than the diameter of your screws)

Sandpaper, medium-fine grit

Water-based polyurethane (oil-based is okay, but I don't like the fumes)

Small paintbrush

2 small brass or stainless steel hinges, ½" x 1½", with eight ½" screws

Scissors

No-see-um netting

Bamboo placemat measuring 11" x 15" (you can purchase these in various sizes, and you'll have to adjust the stretcher bar sizes accordingly)

Needle and thread

1. To make the *keta*, apply wood glue to the joints, and fit them together to form two wooden frames. Make sure the corners are square. Drill holes in the corners, and fit with screws to secure the joints. Sand the surfaces to make them smooth and even.

step 3

step 5

2. Apply polyurethane with a brush to waterproof the frames, and let them dry at least overnight. I recommend applying a second coat.

3. Attach the hinges to one side of a long length of one of the frames, centering them approximately 4" in from each side. Join the two frames by attaching the hinges to the second frame, leaving a ⅛" gap between the two frames.

4. To make the *su,* cut a piece of no-see-um netting to 12" x 17". Lay the netting over the bamboo mat, and cut it to size, making it the exact width and 1" longer than the mat.

5. Wrap ½" of netting around one of the short sides of the bamboo mat, and stitch the netting in place around the wide splint of wood. There should be approximately ½" of netting hanging off the unsewn end of the mat, because the mat will stretch when wet.

To use: When making paper, simply place the *su* on top of the open frame of the *keta,* with the no-see-um netting *sha* facing up. Close the hinged *keta,* sandwiching the *su* tautly in the frame. Hold it together tightly as you form your sheets on its surface. After you make a sheet, open the hinged *keta,* lift the *su,* and gently roll the sheet down onto the couching surface (see Making Eastern-Style Paper in chapter 5).

VARIATION: CONVERTING A WESTERN MOULD INTO A SUGETA

With a few extra materials, you can modify your own Western mould into a *sugeta* suitable for making paper Japanese style.

You Will Need:

Interfacing or fiberglass window screening
No-see-um netting
Duct tape

1. Cut a piece of **interfacing** or fiberglass window screening to the outer dimensions of the mould. You can attach a piece of no-see-um netting to the top of the window screening to give your sheets of paper a smoother surface.
2. Wrap a piece of duct tape around one of the long edges for easy handling (and to connect the two screens if you are using two layers). Make sure the tape sits underneath the deckle and does not block the papermaking surface of the screen.
3. When making paper, simply sandwich this *su* between the Western mould and deckle, and form your sheets on its surface. After you make a sheet, you can lift the *su,* and gently roll the sheet down onto the couching surface (see Making Eastern-Style Paper in chapter 5).

step 2

step 3

Deckle Box

A third type of mould and deckle is called a deckle box. It's designed for pouring sheets of paper, rather than pulling sheets from a vat of pulp. The very first papers may have been made in this fashion on a square of woven cloth held within a bamboo frame. This method of sheet forming is still used today in many countries, including Nepal, Tibet, Thailand, and India.

A deckle box has a deep deckle, which allows you to pour water and pulp into it (simulating a vat) to form a sheet. This type of mould allows you to make paper with a small amount of pulp (not enough to fill a vat). You can also mix various pulps in a deckle box to make a decorative sheet or a work of art. And this setup also lets you make really thick sheets of paper, if you'd like.

Interfacing

a spun or nonwoven polyester mesh often used as a couching surface in papermaking.

MAKING A DECKLE BOX TO CREATE LARGE SHEETS

Abaca *(Musa textilis)*

a bast fiber from the leaf stalk of a type of banana tree found in the Philippines and South America.

Formation aid

a viscous substance that slows drainage and allows time for sheet formation.

TIP

Matchstick bamboo window shades are available at import stores in custom window sizes; buy one that is large enough to cut to 25½" x 37½".

This deckle box design comes from papermaker Andrea Peterson of LaPorte, Indiana. Her design is for sheets that are 24" x 36", but you can construct yours to the size you want. Dipping large sheets of paper can be cumbersome because of your limited arm-span and the weight of the pulp. But by pouring pulp into a deckle box, you can make these large sheets yourself. With this design, the deckle box is also your vat. This technique works best for long fibers, such as Asian fibers and **abaca** mixed with **formation aid.**

You Will Need:

For the top frame (deckle):
2 1 x 3s cut to 25½" lengths, with mitered corners (joint the corners in any fashion; just be sure they're square)
2 1 x 3s cut to 37½" lengths, with mitered corners
Galvanized finishing nails or brass screws
Hammer
Screwdriver
4 strips of 1 x 1 cut to 8" lengths, with mitered corners

For the bottom frame (mould):
2 1 x 2s cut to 25½" lengths, with mitered corners
2 1 x 2s cut to 37½" lengths, with mitered corners
6 strips of 1 x 1 cut to 24" lengths
Polyurethane or wood sealant
Drill
4 door-latch hooks and 4 eyehooks to fit door latches
Matchstick bamboo window shade (see box at left)
Several sheets of 26" x 38" heavyweight interfacing

Important: the bamboo strips must run in the 37½" direction. If you have them the other way, the support structure for the bamboo will not function properly.

1. Construct the top frame of the deckle box using the 1 x 3s, and nail or screw the joints together.

2. Use the 1 x 1 strips to brace the corners of the deckle. Make sure they are flush with the top of the frame; then nail or screw them to the frame.

3. Construct the bottom frame of the deckle box using the 1 x 2s, and nail or screw the joints together.

4. Nail or screw the 1 x 1s to the frame at approximately 5" intervals. Attach them at an angle so that a corner of the wood is flush with the top of the frame.

5. Seal all wooden parts with polyurethane or wood sealant.

6. Drill pilot holes, and then attach two door-latch hooks to each side of the 37½" sides of the top frame, approximately 6" in from the ends and centered on the top frame.

7. Place the bamboo window shade and a piece of interfacing on top of the bottom frame.

8. Place the top frame on the bottom frame assembly, and attach the four eyehooks to the bottom frame. Make sure that when the latches are hooked, the interfacing and bamboo window shade are held securely between the frames.

Preparing the Pulp

For abaca or **kozo,** use 1 pound of dry fiber to make 10 to 15 sheets of paper that are 24" x 36".

1. Beat 1 pound of pulp.

2. Scoop about 20 ounces of pulp into a 24-ounce container; then dump this into a 2-gallon bucket filled with water.

3. Add approximately ½ cup of concentrated formation aid (see recipe on page 74), and mix thoroughly.

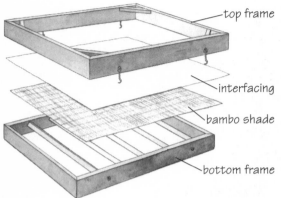

top frame

interfacing

bambo shade

bottom frame

This deckle box design allows you to pour the pulp into the box to make larger sheets of paper.

Kozo

the Japanese name for *Broussonetia kazinoki* or *Broussonetia papyrifera*, two of the many members of the *Moraceae* family, commonly known as mulberry, whose inner bark is used in Japan for papermaking.

MAKING PAPER WITH THE DECKLE BOX

For this particular project, I recommend you work outside or in a space that can get wet, because this is a very wet process!

You Will Need:

Deckle box
Table or two sawhorses
Hose or bucket of water
Plastic sheeting cut to approximately 40" x 50" (a large black garbage bag, cut open, works well and makes it easy for you to see the pulp and gauge whether you have enough)

step 1

1. Place the mould on a table or two sawhorses (if you are working alone, two sawhorses work best: place one a few feet in from the end of the short side of the deckle box, and the other almost at the end of the deckle box).

2. Set the bamboo window shade on top of the mould, followed by a piece of nonwoven interfacing, and make sure everything is smooth and flat.

step 4

3. Soak the entire apparatus with water, either with a hose or by pouring a bucket of water over it. Then place the deckle on top, and fasten the door latches.

4. Cut a piece of plastic sheeting large enough to fit into the deckle, and drape it slightly over the edges.

5. Pour a sheet's worth of pulp into the plastic-lined deckle box (this will take some experimenting and will depend on the thickness you are after, but use the pulp recipe above as a guide). Add pulp or water if necessary.

6. You are now ready to perform the magician's trick of pulling the tablecloth out from underneath the dishes: Take hold of one 36" side of the plastic sheeting, and gently but firmly pull it out from underneath the pulp and water in one fluid motion. Set the plastic sheeting aside. As the water starts to drain through the interfacing, quickly move to the end of the deckle box, which is just resting on the sawhorse. Lift the end of the box, and gently rock it back and forth to distribute the fibers evenly on the mould surface. (You can use the corner braces as handles, or you can install door handles on the ends of the bottom frame at a comfortable distance from each other). If you are working with two people, you can each lift an end and gently shake the sheet. Rest it back on the sawhorse and allow the sheet to drain thoroughly.

7. Undo the door latches and remove the deckle, taking care not to drip onto the freshly made sheet.

8. Have each person take one of the small ends of the paper, lift the interfacing with the sheet of paper, and transfer it to another surface. No couching is necessary in this process; the sheets are formed directly on the felts.

9. Place another piece of interfacing on top of the bamboo window shade, reassemble the deckle box, and continue making sheets.

step 8

Note: You must have a fairly large press to accommodate these sheets of paper. One alternative is to press the sheets individually using a brayer: place a piece of interfacing and a piece of cloth on top of the sheet, and gently roll over the entire sheet's surface. Push the brayer from the center out, to remove water. If the cloth gets really wet, you can change it to absorb more water. After pressing, you can dry the sheets using a variety of methods. Andrea Peterson brushes hers onto smooth, large, melamine-coated boards.

You can use this sheet-forming technique with any size deckle box if you don't want to bother with setting up a vat; but you'll need the door latches to ensure that the interfacing stays in place.

For more information on making large sheets, read *How to Make Big Sheets,* by Edwin Jager (see reading list).

MAKE YOUR OWN DECKLE BOX

You can modify a Western mould and deckle to create a deckle box by constructing an insert for your deckle.

You Will Need:

Materials to fit an 8½" x 11" mould and deckle (adjust sizes for other dimensions)

2 pieces of ¼" or ⅜" thick Styrofoam, cut to 4" x 8½"

2 pieces of ¼" or ⅜" thick Styrofoam, cut to 4" x 10¼"

Waterproof tape

step 1

1. Make a box out of the four pieces of Styrofoam by joining the pieces at the joints with waterproof tape. Make sure that the 11" pieces fit inside the two 8½" pieces (you might want to assemble this unit inside of the deckle for a sure fit).

2. Place the Styrofoam box inside the existing deckle, and tape the two together.

step 2

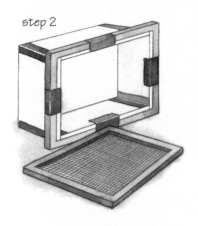

If you don't feel like making your own deckle box, there are a couple of suppliers who carry ones that are a bit more sophisticated. The one disadvantage of the self-constructed box is that the mould and deckle are not attached to each other, which makes sheet-forming slightly cumbersome. Arnold Grummer manufactures a neat deckle box that fastens with Velcro and can be found in many art supply stores, and Jana Pullman creates beautiful deckle boxes that are fastened with latches (see resource guide).

For instructions on making paper using the deckle box, see chapter 5.

WHICH MOULD FOR WHICH PROJECT?

The Western mould and deckle is best for
+ Single-dip sheet-forming process
+ Traditional Western fibers, such as cotton and linen
+ Most preprocessed sheet fibers, such as abaca, hemp, and flax
+ Some plant fibers (if sheet formation is difficult, try using an adapted su or a sugeta)

The sugeta is best for
+ Multiple-dip sheet-forming process
+ Long fibers, such as kozo and other plant fibers
+ Making thin sheets
+ Any fiber that is difficult to couch

The deckle box is best for
+ Pouring rather than dipping sheets of paper
+ Making thick sheets
+ Small quantities of pulp
+ Mixing various pulps together in one sheet of paper
+ Making test sheets (no need to fill up an entire vat just to make one sheet)

VATS

The vat is the container in which you form sheets of paper. This tub gets filled with a slurry of water and pulp, and you dip the mould and deckle into it to form sheets. You can use many types of containers as vats, as long as they don't leak or rust. The vat needs to be at least 5 inches bigger than your mould and deckle (on each side), so that you have room to dip into it. It should also be at least 6 inches deep — more so if you are making large sheets — but it does not need to be deeper than is comfortable for sheet formation. Deep vats require more pulp, and they take longer to fill and empty.

The size of your mould and deckle should determine the size of the vat you choose.

A drainage hole makes a vat easier to empty.

Dishpans or plastic utility tubs make excellent tabletop vats. Free-standing plastic and fiberglass utility sinks without plumbing make perfect standing vats, and they have drain plugs, which make them easy to empty. You can also use a clothes storage bin or a restaurant bussing tub. A really simple vat can be made with a piece of plastic sheeting and a laundry basket. Just line the basket with the plastic, clip the loose ends to the lip of the basket with clothespins, and fill it with water and pulp.

I do not recommend using your sink or bathtub as a vat, because the drains will inevitably clog with fiber.

EASY USE FEATURES

The height of your vat is important; you should be comfortable when making sheets. It's nice to have standing vats on caster rollers, for easy mobility and cleaning. Having a hole that can be plugged on the bottom or side of the vat is handy for emptying: simply slip a bucket and strainer underneath, and pull the plug to drain out the excess fiber and water.

HOSE ATTACHMENTS

In papermaking, you need water all the time — to fill a vat, to rinse cooked fiber, to clean your equipment — so do yourself a favor and invest in a hose with a spray nozzle. Get a hose with a latch so you can lock it in the "on" position when you are filling a vat (just keep an eye on it so it doesn't overflow!). A spray nozzle with varying functions (jet, mist, etc.) can be useful for a variety of techniques: misting colored pulp, creating imagery in a pulp painting, using a full-force jet spray to clean the fibers from the screen surface of the mould.

MEASURING DEVICES

A kitchen scale will come in handy for weighing fiber before cooking or beating, and for measuring the amount of alkali to add to the cook-

TIP

If you use a nonplumbed plastic utility sink with legs as a vat, it can easily be disassembled — and you can store the legs and your papermaking equipment in the tub, which takes up only a small amount of space.

ing solution. You will also need a dedicated set of measuring spoons (not aluminum) for measuring alkali, **pigments,** and other additives.

COUCHING SURFACE

After you make a sheet of paper, you need to couch it — or transfer it — to another surface. You can lay the sheets one on top of the other, with interleaving felts, forming a pile called a post. The surface you use as a couching stand should be waterproof or covered with plastic sheeting, because it will get wet. You might need to protect the floor, too.

A post of paper is very wet and heavy, and it can easily be damaged if it is not properly supported. In most cases, you will need to transport the post of paper to a press or drying area, so you will need to couch onto a portable surface — such as a waterproofed board, or a piece of stainless steel or galvanized metal sheeting — not directly onto your work table. You can also couch into a tray with a lip, such as a cafeteria tray or a baking sheet; this will collect excess water, which you can pour off from time to time. You can even couch directly onto the bottom press board if you are using a press. Some papermakers couch directly onto the drying surface, such as interfacing or boards (see page 34).

WATER CONTROL

Here are a couple of nifty systems for controlling water on a tabletop:

◆ For a temporary setup, place 2 x 4s around the edge of the tabletop, leaving a gap at one corner. Drape a plastic sheet over the top, and create a channel at the gap for the water to flow into. Place a bucket underneath.

◆ Attach pieces of wood, such as 2 x 4s, around the edges of a tabletop, forming a lip that will trap water. Polyurethane the tabletop,

Pigment

an insoluble, finely ground coloring material that can be used to color pulp. Pigments must be attached with a retention agent.

A cart with casters can make it easier to transport wet sheets.

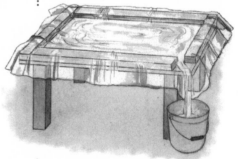

Arrange 2 x 4s under a plastic sheet on a tabletop, leaving room at one corner for a water channel.

and seal the joints with the 1 x 2s to prevent leakage. Drill a hole in one or two corners of the tabletop for the water to drain into, and place buckets underneath to collect the water.

FELTS

With most papermaking styles, you will transfer the sheets to felts as you make them, freeing up the mould and deckle so that you can form more sheets. Traditionally, wool felts were used. Papermaking "felts" are woven, but they have the texture and surface of real felts, which are matted. The term *felt* is often used by papermakers to refer to any couching material. If you use true felts, you will most likely find that they lose their shape and do not hold up over time.

Old woolen army blankets make great felts, and you can often find them at secondhand shops. If you live near a commercial paper mill, you might ask if they have any old commercial felts that you could cut up and use. Nonfusible interfacing, available at fabric stores, is an excellent lightweight material that works well as a felt substitute. It comes in different weights — I like the extra-heavy weight. Papermaking suppliers also carry an assortment of couching materials (see resource guide).

The felts should be cut approximately 2 inches larger than the sheets of paper that will be couched onto them and small enough to fit into your press.

FELT OPTIONS

You can couch your freshly made sheets of paper onto a variety of surfaces, including any of the following:

✦ Nonfusible interfacing
✦ Old bedding
✦ Newspaper (test first, to see if the ink bleeds)
✦ Army blankets
✦ Old felts from a paper mill

Felts can be cut for one sheet of paper, or you can cut them large enough to accommodate several sheets.

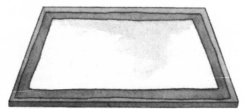

BEATING EQUIPMENT

There are numerous methods of beating pulp, from hand-beating with mallets to using a sophisticated machine called the Hollander beater. Following is a description of each piece of equipment; refer to chapter 3 for instructions on how to beat pulp using these tools.

MALLETS

Mallets are used to hand-beat plant fibers (which have been cooked in an alkali solution or retted). You can use a meat tenderizer, a baseball bat, or a stout stick to pound the pulp. Your mallet should be made from hardwood that will not split or splinter when used. There are also special mallets designed specifically for hand-beating that can be purchased from papermaking suppliers (see resource guide).

BLENDER

Many pulps can be processed in a blender. The blender is the harshest method of beating because it cuts the fiber, rather than merely bruising and separating it, which is how professional pieces of papermaking equipment process pulp. A standard kitchen blender can be used, but I don't recommend using the same blender for food and papermaking. Large commercial-grade blenders are more efficient because they can blend more pulp at a time.

DRILL ATTACHMENTS AND WHIZ MIXERS

Drill attachments for mixing paint work well for breaking down sheet pulps, such as abaca and cotton linters. A contraption called a Whiz Mixer — made just for papermaking — is available from papermaking suppliers. Some papermakers have rigged up systems that use a garbage disposal for recirculating and breaking down pulp. These pieces of equipment can all be used for rehydrating pulp or mixing inclusions or pigments into pulp, but they are not suitable for preparing plant fibers or other raw fibers.

Mallets come in all shapes and sizes; just be sure yours is made of good hardwood.

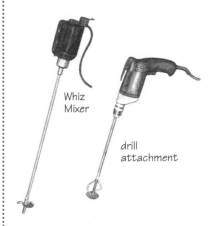

Whiz Mixer

drill attachment

Whiz Mixers and drill attachments are used to rehydrate processed pulps.

Stampers are used in Japanese papermaking.

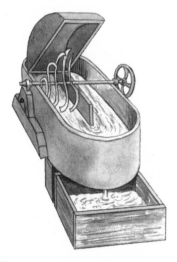

The *naginata* beater has curved knives instead of a bedplate and beater roll.

BALL MILL

Lilian Bell, author of *Plant Fibers for Papermaking,* uses a ball mill for beating many of her plant fibers, particularly long-leaf fibers and some **bast fibers.** The ball mill is similar to a rock polishing machine; it roughens and hydrates the fibers as they move between porcelain pebbles and the sides of a rotating jar. Ball mills are available through many ceramic or lapidary suppliers. They are fairly expensive, beat a small amount of fiber at a time, and are very noisy. You can find instructions for using the ball mill in Bell's book.

STAMPERS

These mechanized systems for pounding fiber duplicate the hand-beating process. Stampers are commonly used in Japanese papermaking, since the long bast fibers need relatively little beating to be separated. A couple of suppliers carry stampers (see resource guide), and Lee Scott McDonald, a papermaking materials supplier, wrote an article about stampers, including detailed instructions for building one, in *Hand Papermaking,* Vol. 3, No. 2 (winter 1998).

NAGINATA BEATER

The *naginata* beater is used in Japan and has rotating knives that do not shorten or cut the fiber. It is used after stamping, to tease bast fibers apart.

HOLLANDER BEATER

The king of papermaking beaters is the Hollander, developed in Holland in the 1680s when paper was first being made by hand in Europe. It is still used today.

Hollander beater

PAPER PRESSES

After a post of paper is created, it should be pressed to remove much of the water from the paper and the interleaving felts. There are a few alternatives to pressing, described on pages 32–33.

There are a variety of simple presses that you can make; instructions for two are included below. If you do decide to make your own, keep the following in mind:

◆ Make sure your press boards are slightly larger than the largest sheets of paper you will make. You can place more than one sheet side by side and build the pile up to utilize the full size of the press.

◆ If you use wood, seal it with polyurethane or any marine varnish so it doesn't mold or warp.

◆ For small presses, an alternative to wood is the plastic material used to make cutting boards. It won't warp, and it's waterproof.

MAKE A C-CLAMP PRESS

You can use any type of C-clamps you might have, as long as they are large enough to fit your post plus two plywood boards.

You Will Need:

    Post of paper ready to be pressed

  2 press boards large enough to accommodate your sheets of paper (see box at right)

  4 C-clamps (or any clamps large enough to fit around the post, including the boards)

Place the post of paper between the press boards, and attach one C-clamp at each corner of the boards. Tighten the clamps a little at a time, starting with one and working around until all four are as tight as you can make them.

Bast fiber

fiber obtained from the inner bark (the phloem, which carries its food and liquids) of plants, located between the outer black bark and the inner woody core.

TIP

A simple substitute for press boards is to use two kitchen cutting boards — they're waterproof, too.

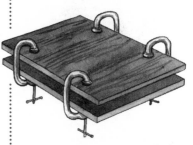

A C-clamp press is simple, but it does the trick!

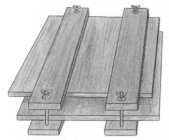

A plywood press is an inexpensive papermaking press.

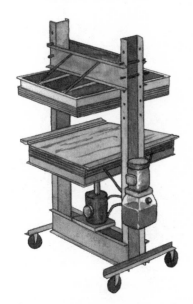

Consider a hydraulic jack press if you want to make a serious commitment to papermaking.

MAKE A SIMPLE PLYWOOD PRESS

With some scraps of plywood, a couple of boards, and simple hardware, you can make your own inexpensive press.

You Will Need:

Post of paper ready to be pressed

2 plywood boards large enough to accommodate your sheets of paper

4 strips of 1 x 4, cut 6" longer than the width of the plywood boards

Drill and ⅜" drill bit

4 carriage bolts, 6" long and ¼" diameter

4 wing nuts to fit the carriage bolts

Drill holes about 1½" in from either end of each 1 x 4, and thread carriage bolts through the holes. Attach a wing nut to each bolt, and screw to tighten the press.

HYDRAULIC JACK PRESS

If you really get serious, this press will ease your processing. A hydraulic jack press allows you to press tall posts of paper and remove a lot of water with very little effort. I have a press with an 8-ton hydraulic jack, which can press about 50 sheets of 22" x 30" paper at a time. Several papermaking suppliers sell presses (see resource guide), and detailed instructions for building a hydraulic press can be found in *Hand Papermaking*, Vol. 3, No. 1 (summer 1998).

FOUND OR ADAPTED PRESSES

You might adapt or use a press that was designed for another use, such as a flower press, cider press, or book press. Just make sure it will accommodate a stack of paper (many printing presses press only one sheet at a time), and watch out for presses with metal parts that will rust.

VACUUM TABLE PRESS

A unique system for pressing uses a vacuum table: a tabletop with small holes in its surface, a holding tank underneath, and a connection for a strong wet/dry vacuum. Plastic is draped over the sheet of paper (covered with a lightweight felt) lying on the table surface. When the vacuum is attached and turned on, it sucks the plastic down onto the sheet and draws water out through the holes in the tabletop.

This system has several advantages. It can press very large sheets of paper, which would be difficult, if not impossible, to fit into a hydraulic press (not to mention lifting them into it). Also, high- and low-relief sheets and sculptures can be pressed in this system. A variety of vacuum systems are sold by papermaking suppliers. If you are ambitious, you could rig up a system of your own.

A vacuum table press has advantages that simpler and less expensive presses cannot offer.

DRYING SYSTEMS

There are many ways to dry sheets of paper. You can air-dry them; you can rig a simple drying rack using a few strips of wood, some nails, and clothespins; you can brush your sheets onto windows or boards; or, if you are really serious about the craft, you can build the efficient drying box system described on page 35.

Flat Surface Drying

You can dry the paper on any clean flat surface that will release the paper when it's dry, such as glass, nonrusting metal, Formica countertops, plaster walls, or Masonite boards. The surface should be clean, dry, and in a well-ventilated area (you can use a fan to circulate air). Keep in mind that you will need enough surface area to accommodate the entire post of paper.

You can brush freshly pressed sheets of paper onto a clean, flat surface, and they will dry flat.

MAKE A DOWEL DRYING RACK

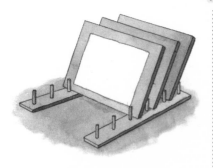

A dowel drying rack is a fun project for a Saturday afternoon.

Patricia Grass, a papermaker and book artist in Portland, Oregon, has a simple but elegant board drying system. She uses a plastic dish rack or letter tray to hold several small boards. She's even gone further and designed this custom drying stand with dowel rods and pine.

Grass recommends birch-faced plywood. Apply polyurethane to one side, to prevent the wood from staining the damp paper. After making and pressing her paper, Grass transfers the sheets to the polyurethaned side of the boards and stacks them in the space-saving rack. This rack will fit five boards and is suitable for sheets up to 8½" x 11".

You Will Need:

2 strips of 12" x 2" x 1" pine
Drill and ¼" drill bit
Wood glue
12 pieces of ¼" dowel rods, cut to 4" lengths
¼" polyurethaned birch-faced plywood, cut to 12" x 12" boards

1. Drill six evenly spaced, ¼" holes down the middle of each piece of pine. Glue a 4" piece of dowel in each hole.

2. Set the two stands parallel to each other about 10" apart. Stand the boards in the rack and place a fan nearby to circulate air.

Air Drying

Drying your papers on a clothesline works best when you couch the sheets onto a thin couching material, such as interfacing. After pressing, simply pin the couching material (with the paper on it) to the line. Be aware that dirt and bugs might stick to your paper if it is drying outside.) If the sheets cockle (warp), you can mist and press the sheets to flatten them. Jana Pullman manufactures a rack designed to air-dry papers (see resource guide).

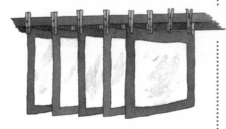

Air dry papers on a clothesline. Use clothespins with a spring; it's much easier to make sure they don't damage the paper.

MAKE A DRYING BOX

Several papermaking suppliers manufacture very efficient drying boxes. I recommend purchasing or constructing a system like this only if you are interested in production papermaking and plan to produce a large quantity of sheets on a regular basis.

This drying system was developed by Andrea Peterson and Jon Hook, a papermaker and a potter who live in LaPorte, Indiana. After working with various drying methods at different studios, they perfected this system, which will dry a large number of sheets very flat. Best of all, it can be broken down easily for storage or transport. I recently built the following one to dry sheets of paper up to 18" x 24".

You Will Need:

¾" plywood cut to 28" x 34" (base of drying box)

22" x 4" x 34" boards

Drill and ⅜" drill bit

4 ⅜" threads with 1" eyelet heads (with nuts) large enough for the tie-down hooks to slip into

22 x 4s, cut to 25" long, with the ends mitered

2 tie-down straps with ratchet tighteners (see box at right)

25 sheets of triwall corrugated cardboard (3 sheets of cardboard that are laminated together), cut to 20" x 25" (the holes in the corrugated must run in the direction in which you want air to flow — the holes on my system run in the 20" direction)

50 pieces of aluminum screening cut to 19" x 24" (optional, but will extend the life of your triwall)

100 blotters cut to 19" x 24" (I purchased blotters that were 24" x 38" and cut them in half)

20" box fan

¾" plywood cut to 20" x 25" (top of drying box)

TIP

Tie-down straps are sold in hardware stores for lashing freight shipped on trucks; make sure you get a ratchet mechanism for tightening, because you will need a secure hold.

If you're planning to dry a large number of sheets of paper and want them absolutely flat, build yourself a drying box.

To construct the drying system

1. Place the two unmitered 2 x 4s underneath the plywood base of the drying box approximately 6" in from one end and 10 inches in from the other end.

2. Drill holes large enough for the eyelets to pass through in the plywood and the 2 x 4s. Center the hole in the 2 x 4, and drill it 2¼" from the end. Repeat on the other side of the 2 x 4 and on both sides of the other 2 x 4.

3. Drill or chisel a hole on the undersides of the 2 x 4s so that the eyelet nut is inset into the 2 x 4.

4. Attach eyelets and nuts.

5. Miter the ends of the two 2" x 4" x 25" boards. (If you wish, you can screw the tie-downs onto the top of the 2 x 4s, which will give you more leverage when tightening).

To assemble for drying paper

1. Place the plywood base (with the attached 2 x 4s and eyelets) on a flat surface. Place the fan at the back of the plywood base, facing toward the front.

2. Place a piece of triwall on top of the bottom board, right in front of the fan, centering it left to right and lining it up with the front edge.

3. Place a piece of aluminum screening and then two blotters on top of the laminated cardboard (the blotters should be no larger than the cardboard — two layers protect the cardboard from getting too wet).

4. Place one pressed sheet that is 18" x 24", or as many smaller sheets of pressed paper as you can fit side by side, on top of the blotter.

5. Put another two blotters and then a sheet of aluminum screening on top, followed by another cardboard.

6. Repeat until you run out of paper or your stack is 20" high. Make sure you stack the system evenly, and do not stack it higher than 20".

7. Once you've completed the stack, place the 20" x 25" piece of plywood on top of the stacks. Place the fan on the base board, behind the stack. Cover the top and sides of the system with a piece of plastic sheeting, leaving the front and back ends open. This directs the airflow from the fan through the stack and maximizes the efficiency of the system. Tape the plastic sheeting to the fan so the air will be forced forward. Place a sheet of cardboard or a board against the two sides to keep them flat against the edges of the stack.

8. Place the two mitered 2 x 4s on top of the stack, lining them up with the eyelets. Attach the tie-down straps to the eyelets, place them over the mitered 2 x 4s, and tighten them as firmly as possible. Since most tie-down straps are designed to span the width of a truck, you might want to trim off some of the excess strapping.

9. Turn the fan on high, and check the sheets in 24 hours. When the sheets are dry, unload the system.

OTHER MISCELLANEOUS EQUIPMENT

There are a few other things you'll need to make your papermaking operation run smoothly.

BUCKETS AND CONTAINERS
Containers varying in size from small yogurt containers up to 30-gallon trash cans come in handy for a variety of functions, including storing pulp, coloring pulp, and scooping water or pulp into or out of a vat.

SCOOPS
These are my favorite tools! A handy scoop can be made from an old plastic detergent or milk jug. Just cut the bottom out of the jug and leave the cap on. Turn it upside down, and you'll have a handle to scoop pulp with. These are great for filling and emptying vats of pulp.

TRANSPORTING HEAVY PULP
Since wet pulp and damp sheets of paper are heavy, you might want to install casters on your tables and buckets, so they are easy to move around. Check your hardware store for small dollies designed specifically for maneuvering containers.

Raid the recycling bin for containers and scoop material.

Strainers can be jerry-rigged with materials you have at home, or you can buy professionally designed strainers.

STRAINERS

At several stages in the process, you will need to strain the pulp. After cooking, it will need to be drained and rinsed; after beating you may need to strain it; and when you've finished making sheets, you'll need to strain the leftover pulp out of the vat. Pasta strainers or colanders will do the job, but they tend to allow a lot of pulp to escape. Line these types of strainers with plastic mesh or no-see-um netting to catch all the fibers and prevent clogging the drain and wasting fiber. There are also commercial-grade cone-shaped cooking strainers that have a fine mesh and conveniently fit into buckets; these are available from papermaking or restaurant suppliers.

STORAGE SPACE

Inevitably, you will end up with stacks of paper that need a home. Flat files, large drawers, or shelves can easily serve that purpose. Archival storage boxes work well too, as do large, flat boxes made for storing photographs and film.

Here are a few other items that are handy to have in your studio:

♦ Sponges for cleanup or even pressing the sheets
♦ Mop and/or squeegee to keep the floor dry and clean up pulp spills
♦ Resealable plastic bags for storing leftover pulp in the refrigerator
♦ Plastic drop cloths or other plastic sheeting for keeping all surfaces dry

CHAPTER 2

COLLECTING AND PREPARING THE FIBER

Paper can be made out of a variety of materials, ranging from recycled paper to plant fibers, such as flax, corn husks, or iris leaves, to preprocessed pulps, such as, cotton linter and other natural fibers including cotton and linen rags. All these materials contain fiber, an essential ingredient in papermaking. Fibrous plants have been used throughout history to create many of the things we use every day, such as clothing, linens, baskets, and rugs.

Fiber comes from plants with an abundant supply of **cellulose.** All plants contain cellulose, but some contain a higher percentage than others. Cotton linter, one of the staples in Western papermaking, contains the most cellulose — up to 95 percent.

The following guide to various materials includes those that can be processed easily with simple household equipment, as well as those that require special tools.

Cellulose

long-chain polymers of sugars, which are the fundamental ingredients in paper.

RECYCLE IT!

Recycling paper is easy — just think of all the junk mail, notes, and letters you discard in any given week. You can repulp these to create new and interesting handmade papers. Turn junk mail into writing papers, turn office waste into cards and envelopes and recycle those old love letters into heart-shaped sheets.

Preparing the Material

Collect at least 1 pound (dry weight) of paper, and tear or cut it into 1- or 2-inch squares. Separate the paper into batches by color if you wish to make various colored papers (for example, use legal pad paper to make yellow paper). Avoid glossy printed papers, such as pages from a magazine, because they are full of ink and other coatings that make them more difficult to break down. Tissue paper and newspapers will work, although they are weaker and will result in fragile sheets. You can always combine papers — try adding some colored tissue paper to white paper to color the pulp, or combine white paper and newspaper to get a mottled effect. You can even make paper from brown paper bags.

Once you have collected and shredded your paper, proceed to beating the pulp (see chapter 3). You can successfully beat recycled paper in a blender, or you can use any of the pulp reconstitutors or even a Hollander beater.

Feel free to experiment with combinations of colors in the papers you choose for your pulp.

BUY IT PREMADE

A wide variety of preprocessed fibers can be purchased from papermaking suppliers (see resource guide). The following pulps can be rehydrated and processed in a blender or a variety of other machines designed to break down pulp (see processing fiber, page 48). You should be aware that the paper made from preprocessed fibers will look different from sheets made from the same raw plant material because of the many variables in the processing of the fibers.

Common Preprocessed Fibers

Generally, you process these materials in the same way you process recycled paper. Cut or tear the sheets into small squares (if they are tough to tear, such as abaca, dunking them in water first will make them easier to tear). These pulps can all be processed in a blender or with a paint mixer or Whiz Mixer, but if you wish to produce crisp and translucent sheets of paper, you will need to use a Hollander beater.

Many of these fibers are also available in their raw, unprocessed forms, but that requires more work on your part.

ABACA (the Philippine word for Manila hemp)

Abaca is the leaf-stalk fiber of a type of banana plant *(Musa textilis)* that grows in the Philippines. It's a versatile, long-fibered pulp that produces strong sheets of paper. It can be beaten in a blender or other pulp reconstitutor, but its real potential is realized when it is beaten in a Hollander beater. The longer abaca is beaten, the more translucent the resulting sheets become. Overbeating also promotes shrinkage, which makes it a good pulp for sculptural techniques. Abaca is also a good substitute for Japanese-style papermaking, since it is a long fiber, and it doesn't require hand-beating. It comes in a variety of grades: bleached, unbleached, and natural.

COTTON LINTER

After cotton seeds have been ginned for textile use, they are ginned again several times to collect any excess fiber. These short cotton fibers, which are not strong enough for textiles, are made into sheet pulp for papermaking. Cotton linter is inexpensive and versatile. Its short fibers make it good for casting, and it also makes an adequate drawing, printmaking, and watercolor paper. It comes in various grades, ranging in length of fiber: first cut is the longest (from the first ginning), and third cut is the shortest.

THE TRUTH ABOUT LINTERS

Many people confuse the word *linter* with an actual sheet of pre-processed pulp. However, it is a term specific to cotton (so there's no such thing as an abaca linter) that is processed in a machine called a linter. The linter is what removes the seed hairs from the seed.

HEMP

Hemp comes from a variety of the *Cannabis* plant. Like abaca and flax, it's a long and strong fiber. It is thought to have been used to make paper in China 2,000 years ago.

SISAL

This tough, long, stringy fiber is extracted from the plant *Agave sisalana*. This is a long-fibered pulp, similar to abaca, but both the raw material and the paper made from it are coarser in appearance.

FLAX

Linen cloth is manufactured from this plant. Flax is a very strong fiber, and its qualities include translucency and high shrinkage when it is beaten for many hours in a Hollander beater. "Bleached flax" comes in sheet form, and although it can be hydrated in a blender or mixer, its real potential is better achieved in a Hollander beater.

OTHER OPTIONS

Many other fibers and pulps are available from suppliers, such as esparto grass, jute, and raw cotton. Check the resource guide, and inquire with suppliers if you are looking for a particular fiber.

Rag Paper

If you want to make rag paper, you'll need access to a Hollander beater, because a blender isn't strong enough to loosen tightly woven cloth. In addition, you'll need to cut the rags into 1-inch square pieces before beating. Cutting rags by hand is an enormous project. I've seen papermakers use pattern cutters — electrically powered spinning blades — and paper shredders to cut their rags. A quilter's rotary cutter could also aid in cutting fabric. I also recommend cooking the rags in a mild alkali solution before beating (see page 56).

RAGS TO RICHES

When papers were first being made by hand in Europe, rags became widely used as a raw material. Until the late 17th century, cotton and linen rags were fermented so that they could be broken down into a usable pulp with stampers. Then the Hollander beater was developed, and it provided a much improved method of processing fiber into pulp.

According to paper historian Dard Hunter, there was once such a rag shortage that people were required to save their rags. "To encourage the use of wool and at the same time save linen and cotton for the papermakers, the English Parliament in 1666 decreed that only wool could be used in burying the dead." In one year, Hunter claims, approximately 200,000 pounds of linen and cotton were saved for the papermakers by this edict.

PICK THE PLANTS YOURSELF

To make paper from plant fibers, you can use just about anything from iris and gladioli leaves to onion skins, leek tops, and corn husks. Some plant fibers must be extracted (steamed, scraped, stripped, etc.); most require cooking; and all must be literally beaten to a pulp. There are a multitude of ways to prepare each pulp, and each variation in the processing will affect the appearance, opacity, strength, durability, longevity, color, and texture of the resulting paper.

Plant fibers are generally separated into three categories: bast, leaf, and grass fibers.

Harvest plants for pulp from your vegetable or flower gardens and even the trees in your backyard.

Bast Fiber

Bast fiber — the inner bark of tree branches and the stalks of herbaceous annuals and perennials — is located between the outer bark and the core of a branch. The inner barks of many plants are quite fleshy and yield a good amount of fiber for making strong, lustrous sheets of paper.

Bast fibers were used to make the first papers in Asia and are still in use today. Common Asian woody bast fiber plants include kozo (*Broussonetia papyrifera*), mitsumata (*Edgeworthia papyrifera*), and gampi (*Wikstroemia retusa*). Common Western herbaceous bast fibers include hemp and flax. Other bast fibers that grow in North America include willow (*Salix* spp.) and paper mulberry (the same plant as kozo).

There are two main types of bast fiber: woody bast, from the branches of trees such as elm (*Ulmus americana*) and paper mulberry (*Broussonetia papyrifera*); and herbaceous bast, from the stalks of herbaceous annuals and perennials such as milkweed (*Asclepias speciosa*) and nettles (*Urtica dioica*). A third type of bast fiber is found in the petiole, or leaf stem, of plants such as abaca (*Musa textilis*). Petiole fibers are typically long, tough, and stringy — similar to the stringy fiber found in pineapple tops.

Woody bast comes from tree branches.

To obtain the usable fiber in bast fiber plants, the branches or stalks must be steamed (to separate the inner bark fiber from the core) and then scraped (to remove the outer dark-colored bark). If you purchase the Asian bast fibers from papermaking suppliers, they come already steamed, stripped, and scraped.

If you cut a shoot of a tree, shrub, or vine and look at its cross-section, you will find the bast fiber (inner bark) between the outer dark-colored bark and the core. To collect the woody bast fiber, choose shoots or branches of trees that are ½ to 1 inch in diameter.

Smaller shoots will have less fiber, and larger shoots might be tough to process. Try to pick shoots that are similar in width, because they will cook more evenly. Cut the shoots at a 45° angle near the base or just above a bud, leaving the main plant intact so that it can continue to grow. The angle is important, because it will aid later in the stripping process.

Many papermakers harvest bast fibers while they are pruning. Remember that you use only the inner bark for papermaking, so you will need a fair amount of branches to produce a small amount of paper. Harvest at least five or six branches that are each about 5 to 6 feet long. Strip the leaves and twigs from the branches.

Herbaceous bast is collected in a similar fashion to the bast of trees, by cutting the stalks at an angle. Annuals such as hollyhock (*Alcea rosea*) and okra (*Hibiscus esculentus*) can be pulled up by the roots. The roots, leaves, and twigs should be removed before you strip the bark to get to the bast fiber.

Be sure to cut the shoots at a 45° angle to make them easier to strip later.

THE RICE PAPER MISNOMER

Rice paper is a term that has been used to denote all Asian papers. This is misleading because it implies that Asian papers are made from rice; in fact, they are made primarily from three different trees that are native to Asia: kozo, gampi, and mitsumata.

Another paperlike material called rice paper is actually made from the pith (the core of the stem) of the rice paper plant, kung-shu (*Tetrapanax papyriferum*), which grows in the swampy forests of Taiwan. This "rice paper" is made by rolling the pith against a knife on a hard, flat surface and then cutting it into thin sheets. It has been used as a surface for watercolor paintings (and, believe it or not, in manufacturing helmets!). Since this product is made from the rice paper plant, the term rice paper is more accurate in this case. Technically, however, true paper is made from a beaten pulp that is then formed into sheets on a screen.

Petiole bast comes from the stems that support the leaves and are connected to the plant's stalk. In manila hemp (abaca) plants and many palms, the leaf stems are often many feet in length and contain long strands of bast fiber that are easily obtained once the stems are cut from the plant. Generally, this type of bast fiber is easy to harvest. Just cut the leaf from the trunk of the plant, and then peel or scrape the skin from the fiber. In some cases, cooking or retting (fermenting) will be necessary to remove the skin.

Leaf Fiber

Long, flexible leaves from plants like iris (*Iris* spp.), lilies, and yucca *(Yucca filamentosa)* yield suitable papermaking fiber. Leaf fibers such as daylily *(Hemerocallis)* are soft and flexible, and the entire leaf can be used, whereas fibers such as mother-in-law's tongue *(Sansevieria trifasciata)* are tough, and the strong, stringy fiber must be isolated from the fleshy material encasing them. Although separating the fiber is a labor-intensive process, these types of leaf fibers tend to yield a whiter, stronger paper.

Collect at least one dry pound of leaf fiber, either by cutting fresh green leaves from the plant in late spring after it has flowered, or by pulling the leaves in the fall when they are brown and wilted (the latter is better for the plant). If you are harvesting live, green fiber, collect more of it, because the fleshy material will disintegrate during cooking, leaving you with less plant material.

Some leaves can be cooked immediately; others must first be scraped or retted to remove unwanted fleshy material. If you are using soft, flexible leaves (such as daylily), you can cook the fresh, green leaves right away or dry them first. For tougher leaves, you'll need to ret or scrape them to extract the fiber before cooking. See pages 49–51 for instructions on retting and scraping.

The entire leaves of some plants can be used in the pulp; with others you must separate the fiber from the flesh.

You can wait until plant leaves are brown and wilted in the fall before pulling them.

Grass Fiber

You can make paper with grasses such as wheat straw *(Triticum aestivum)*, feather grass *(Phragmites communis),* and corn husks and stalks *(Zea mays).* Most grasses do not require further processing prior to cooking.

Harvest about a pound (dry weight) of fiber. You can harvest grasses when they are fresh or dry, either by pulling them up by the roots or cutting them. Grass fibers are the easiest to process, but they usually break down significantly, yielding the least amount of pulp.

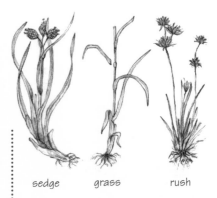

sedge   grass   rush

Other Plant Fibers

Plants you might expect to make the most beautiful papers might not contain enough fiber to hold together as sheets. For example, flower petals and tree leaves will not produce sheets by themselves. However, these types of plant material can be used decoratively when tossed into a vat of a more substantial pulp.

THE POSSIBILITIES ARE ENDLESS

You can make paper from any number of plant fibers found in the countryside, your garden, or even your kitchen compost. Here are some suggestions.

Kitchen Castaways
- Artichoke leaves
- Broccoli stalks
- Corn husks
- Leek greens
- Onion and garlic skins
- Rhubarb ends

Garden Snippings
- Canna lily leaves
- Daylily leaves
- Hollyhock stalks
- Hosta leaves
- Iris leaves
- Yucca leaves

Wild Ideas
- Cattail stalks
- Hay
- Milkweed
- Nettle
- Thistle
- Wheat straw

cutaway view

Steam your cut stalks in a pot covered with another pot to accommodate their height.

When you can see the bark shrinking from the end of the wood, the stalks are ready to strip.

PROCESSING TECHNIQUES FOR PLANT FIBERS

All plant fibers require processing before they can be beaten. Some plant fibers need to be extracted from the plant; others need to ferment; and almost all require cooking before they can be beaten. Purchased and recycled fibers have already gone through these types of processing to get to their current state.

Steaming

The first step in processing a bast fiber (excluding petiole bast, which can usually be peeled right from the plant and does not require further processing) is to separate the inner bark (bast) from the outer bark and core, if there is one. With some plants — such as willow, elm, and gampi — this can actually be done right after harvesting, but most plants require steaming to facilitate removal.

Place the cut stalks in a large covered pot, leaving them as long as you can. If you have two pots that are approximately the same size, you can use one as a lid by putting it upside down on top of the other, thus allowing for longer stalk lengths. Otherwise, you can make an aluminum foil tent to cover the stalks. Put approximately 2 inches of water in the bottom of the pot, and place the stalks in vertically. Cover tightly with a lid or the foil. Bring the water to a boil, and steam the fiber until you see the bark shrinking from the end of the wood — this will be obvious to see and can take anywhere from half an hour to 2 hours. Turn off the heat.

Stripping

As soon as you can handle the bark, remove the branches, and strip the inner bark from the core of the plant while it is still warm. To strip the inner bark, take a branch and make a vertical slit at the bottom of the stalk (the wider angled end). Pull enough away from the core to get the entire stalk's inner and outer bark in your hand. Note that stripping from the wider base toward the end of the branch (large end to small end) will aid in removing the entire inner bark in one pull. Hold the large end firmly, and pull the bark off the core at a severe angle.

Cleaning

If the outer bark was not removed during the stripping process, you will need to scrape it off the inner bark. Even if the outer bark fell off during stripping, there is probably some remaining bark that could be removed. Scraping off the outer bark ensures that your paper will turn out pure and consistent, because the fibers can be properly separated during beating.

You may see several layers of inner bark in some plants. Paper mulberry, for example, has both a white and a green inner bark. The white is purest, and you can separate it from the green bark to make the finest paper — or combine it with the green bark, which is also fibrous. The inner bark must be wet to facilitate scraping. If it's dry, soak it in water for at least an hour. Take one piece of bark at a time, and with a blunt knife or other scraping tool, scrape the fiber against a hard surface. Start at the wider end of the bark, and scrape from the base toward the top.

Wash the stripped bark in clear water, and pick out any remaining specks of bark. At this point you may proceed with cooking the fiber, or you may hang it to dry (see Drying, page 51), then bundle and store the fiber until you are ready to make paper.

Retting

Tough leaf fibers (such as yucca) and herbaceous bast fibers (such as milkweed) are difficult to break down just by cooking. Overcooking the fiber or cooking it in too strong a solution can destroy its

Holding larger branches between your legs will give you good leverage to strip both the inner and outer bark in one pull.

Before you begin the cooking process, you will need to strip any remaining outer bark off the inner bark.

integrity by breaking down the cellulose. Retting (fermenting) is a better alternative. Retting accomplishes two things: it loosens the cellulose, and it breaks down noncellulose material, reducing or eliminating the need to cook the fiber. A way of letting nature help you prepare your fiber, it can be as simple as leaving your plants out in the back yard or as complex as dissolving them in an alkaline solution and monitoring the pH. (For more information on retting, see the reading list.)

ENZYME RETTING

Florida papermaker Betty Kjelson uses an enzyme as a catalyst for retting her plant fibers. These fibers include subtropical plants such as banana, palm, sansevieria, and hibiscus. She fills 30-gallon garbage barrels with harvested plant fibers, covers them with water, adds the enzyme powder, and then seals the lid tight. The enzyme powder speeds up the retting process.

Various types of enzyme powder can be purchased at hardware stores. They typically consist of bacteria cultures, enzymes, micronutrients, and inert ingredients. The one Betty uses is normally used to prevent septic system backup. She adds about 2 cups of enzyme powder to 30 gallons of fiber and water. The ratio of fiber to water varies, but the fiber should be loose enough to allow the solution to work on all the fibers.

The garbage barrels sit against a south-facing cement wall, which is so hot that the heat of the sun almost cooks the fiber. Retting time varies from several weeks to several months, depending on the fiber. Test the fiber weekly. When done, it should appear much as a fiber does after it is cooked in soda ash.

After retting, Betty rinses the fiber thoroughly and cooks it in a light solution of soda ash. She then rinses it again before beating it.

Scraping

Leaf fibers such as yucca, sisal, agave, and pineapple, with their long fishing-line–like consistency, are difficult to process and might require the use of a Hollander beater. The stringy fibers must first be separated from the fleshy green leaf matter in a process called decortication. One way to do this is to scrape the fleshy material off with a knife. I've found that pounding the leaves with a meat tenderizer and soaking them in water for a couple of hours before scraping can ease the process.

Work on a hard, clean surface that isn't too slippery (a cutting board is okay), and scrape one leaf at a time. Start at the tip of the leaf, and scrape in 3-inch sections (I've found, in trying to scrape an entire leaf, that the strands get tangled). After removing the fleshy material from the top 3 inches, cut the scraped strings off the leaf and start a new section. This process can be tedious, and if you are interested in producing quantities of this type of paper, you will probably want to ret the fiber first, or find a mechanical device that can aid in decortication.

Once you have removed the stringy fibers, proceed to cooking, as described below.

Drying

If you do not plan to make paper with your fiber right away, you can dry and store it before cooking. You should store plant fiber only when it has been processed right up to the cooking stage. Bast fibers should be steamed, stripped, and scraped before storing, and tough leaf fibers should be retted or scraped. It is very important to prevent mold and mildew when drying and storing, so that you do not damage the fiber. Do not store fiber in plastic bags, which trap moisture. To dry fiber, hang it or lay it out on a screen so air can circulate all around it. If you wish to whiten or lighten your fiber, try drying it

When you separate the fleshy part of the leaf from the stringy fibers, work in 3-inch sections to avoid getting the strands tangled.

As you finish scraping each 3-inch section, cut it off and start a new one.

outside and letting the sun bleach it. Once the fiber has dried, tie it in bundles for storage, and label it. If you must bag your fiber, put it in paper, mesh, or cloth sacks.

COOKING BASICS

Just as most plant fibers require some processing before they can be cooked, almost all plant fibers need to be cooked before beating (retting is an alternative with some fibers — see page 49). With the aid of an alkali, cooking softens tough, raw fiber to facilitate subsequent beating. In addition, cooking removes noncellulose materials such as **lignins,** pectins, waxes, and gums, which — if not removed — interfere with the bonding process and will affect the archival qualities of the paper. During cooking, they are dissolved and then rinsed out.

Cooking Equipment

For cooking plant fiber, you'll need stirring utensils and pots that you do not use for cooking food. Do not use aluminum or iron pots, because the alkali used in cooking the fibers will react with those materials and produce a gas. Stainless steel pots are ideal, but you can also use glass or enamel-coated ones. Also use stainless steel, wood, or heat-resistant plastic utensils.

It is best to cook fiber in an area other than your kitchen, if possible, so that you don't contaminate any food or food preparation surfaces. I use a camping stove outside. Your cooking area should also be well ventilated, to help dissipate fumes.

Breaking Down the Fiber

Several alkalis can be used to break down plant fibers: soda ash (Na_2CO_3); washing soda; lye (NaOH, sodium hydroxide, caustic soda); wood ash (KOH); and lime ($CaOH_2$, calcium hydroxide). The most

Lignin

the cementing material between the cellulose and cell walls in plants, which can contribute to premature deterioration and discoloration in paper.

In choosing cooking equipment and in siting your cooking area, you need to consider safety first.

common alkali used by papermakers is soda ash, which is available from papermaking and ceramic suppliers. (For instructions on using other alkalis, please refer to *Papermaking with Plants, Japanese Papermaking,* or *Plant Fibers for Papermaking,* on the reading list.)

You should cook the fiber in the least amount of soda ash needed to break it down. Trying to speed up the process by increasing the amount of soda ash could damage the fiber. You can adjust the ratio as you gain experience, adding less soda ash for weaker fibers and more for tough fibers. Keep a record of how you prepare certain fibers to help you gauge the correct proportions.

TO COOK OR NOT TO COOK

You might try cooking recycled paper, to remove ink, and rags, to remove any starches or other additives that were intended to enhance the life of the fabric.

SAFETY

Although the likelihood of an accident is rare because of the small amounts of alkali you will be using, there are a few safety measures to follow. Soda ash, and other alkalis, can burn if they come into contact with skin. Be sure to use caution when working with any alkali.

+ Never use aluminum, tin, or iron pots, which react with the alkali and produce a gas. Instead, use stainless steel, enamel-coated, or glass pots.
+ Use stainless steel, heat-resistant plastic, glass, or wood utensils and tongs.
+ Keep your cooking utensils and pots separate from your papermaking utensils and pots. If possible, do not work in your kitchen.
+ Use a hooded/vented stove to protect yourself from cooking vapors. If this isn't possible, cook outside with a camping stove.
+ Add alkali to water before it boils. Do not add boiling water to alkali or vice versa; it could splatter and burn you.
+ Wear rubber gloves when working with alkali.

Basic Cooking Process Using Dry Weights

This process will take approximately 2 hours, but it might need to be adjusted for some fibers. If you have more or less than a pound of dry fiber, simply divide or multiply using the formula below. Keeping records will simplify your processing in the future.

You Will Need:

step 1

1 pound of fiber
Water
Scissors (optional)
Cooking pot
Rubber gloves

3½ ounces of soda ash
(about 20 percent
of the dry weight of
the fiber)
Tongs

1. Weigh out the dry fiber on a kitchen scale. Be sure to weigh the fiber before wetting it — measurements given are based on dry fiber weights.
2. Soak the fiber in plain water for a few hours or overnight. If it floats on top of the water, weight it down with a bowl or some other heavy object to ensure that all the fiber absorbs some water. This swells the cell walls of the fibers and makes them more receptive to cooking.

If you plan to beat the fiber in a blender or a Hollander beater, you should cut it into ½" strips before soaking it, to prevent tangling. You can do this with a pair of scissors, although I find that my hand tires quickly. I have an old paper cutter, which I use to chop my fiber; other papermakers use garden shredders.
3. Fill a nonreactive cooking pot with enough water to cover the soaked fiber, but do not add the fiber yet. Bring the water to a boil.

step 4

4. Wearing rubber gloves, add the soda ash to the water before it boils and before you add the fiber. After the soda ash dissolves (a matter of seconds), add the soaked fiber, and stir it into the pot. Turn the heat down, keeping it at a simmer.
5. Every 30 minutes, stir the fiber in the pot, and test it for doneness. To test, remove a piece of fiber from the pot, rinse it, and pull it in

the direction of the plant's growth. The fiber is done when it pulls apart easily (see box at right).

6. Turn off the heat, and remove the pot from the stove. Pour the cooked fiber into a strainer, getting rid of the water and plant "liquor" (all that has cooked out of the fiber). Rinse the fiber thoroughly, until the water running through it is clear. It is very important to remove all the soda ash at this point, so that further breakdown of the fiber does not occur. The fiber is now ready for beating.

Cooking Fiber of Unknown Weight

If you harvest green plant material and don't want to dry it — or if you just forget to weigh the dry fiber — try this process, which assumes that your fiber is already damp.

You Will Need:
> Fiber
> Water
> Soda ash

1. Fill a nonreactive cooking pot with water sufficient to cover the fiber and allow it to move freely, but do not add the fiber yet. Measure the water as you add it to the pot, so you'll know how much you used.

2. Heat the water, and before it boils, add ½ ounce soda ash per quart of water (wear your rubber gloves!). When the soda ash is completely dissolved and the water boils, add the fiber to the solution, piece by piece.

3. Reduce to a simmer and cook, stirring the fiber every 30 minutes, until it pulls apart easily (see box at right). At this point you can let the fiber stand in the cooking solution, or you can proceed to rinsing (see step 6 in recipe above).

**PULP TEST:
IS IT COOKED?**

A papermaker I know learned this method for testing fiber for doneness while studying in Korea (there are similar practices in many countries): Grab the fiber with your thumbs and forefingers, with thumb tips touching. Using only the force exerted when you push the first thumb knuckles together, try to tear the fiber apart in the direction of the plant's growth. When the fiber tears easily (it will be somewhat like cooked celery), it has finished cooking.

RINSING VARIATION

Rinsing can be a tedious process. This system, devised by paper-maker Lynn Amlie, works well if you have the time. Drill holes around the top of a 5-gallon bucket, just below the lip. Put the end of a garden hose in the bottom of the bucket, place the fiber in the bucket, and turn the hose on to a slow trickle. When the bucket fills with water, the water will start to drain out of the holes, but the fiber will stay in the bucket. In some cases, small impurities will float away too. If this is done outside, the sun will also whiten the fiber. Since the water is running so slowly, it can take up to 3 hours to rinse completely. When rinsing is complete, the water should not feel slimy, and it should run clear.

Sizing

a substance that makes a paper more water-repellent. Internal sizings, such as Hercon 70, are added to pulp at the end of the beating cycle and are in the paper as it is formed. External, or surface sizing, such as gelatin, is applied to the surface of dry sheets.

Cooking Rags

If you use new rags or textiles, it is a good idea to cook them to soften the rags and remove any **sizing** or finishing materials, such as starches, fire retardants, and waterproofing agents. These materials prevent the fibers from opening up during beating, which means your beating time will be long! Use the basic cooking recipe on page 54, but lower the percentage of soda ash to 15 percent. Thoroughly rinse the rags after cooking. Old rags or clothing do not need to be cooked — wear and tear and washing have removed unwanted materials, and the fabric is usually soft enough to break down quickly.

PAPERWHITES PAPER

There are many ways to process plant fibers. I do think it is helpful to include an example, though, to illustrate how one papermaker produces her paper. Just remember that this recipe may not work for all plant fibers. In my book *Papermaking with Plants,* I've included a recipe for a plant fiber in each of the fiber categories.

Papermaker Kathy Crump of Stockton, California, has been making paper from paperwhites *(Narcissus tazatta),* and she currently has about 400 bulbs growing in her garden — enough to supply her with paper for her projects.

You Will Need:
- 1½ pounds dry fiber
- ¾ cup soda ash
- 2 gallons water

1. Fill a 5-gallon nonreactive pot with 2 gallons of water, and bring to a boil. Wearing rubber gloves, measure out the soda ash and slowly add it to 1 quart of warm water in a plastic bucket. Stir with a wooden dowel or stick to dissolve the soda ash, being careful to avoid splashing (soda ash can burn you). Add the soda ash solution to the pot of boiling water, and stir well.

2. Add the dry fiber, stirring it to thoroughly wet it and make it sink into the solution. Cover the pot, bring it to a boil, and then reduce the heat and simmer for 1 hour.

3. Remove a strand of fiber with a wooden dowel or stick, and perform the pulp test (see page 55). If the fiber is not done, continue to cook it, testing every 15 minutes.

4. Drain the fiber as soon as it is done, and rinse it in water until the water runs clear. Then, soak the rinsed fiber in clean water overnight.

5. Drain and soak the fiber in clean water two more times. This extra rinsing is worth it — you will get a brighter paper.

GROWING AND HARVESTING THE PLANTS

To produce the lightest-colored fiber for paper, Kathy covers the ground around her plants with black plastic as the plants grow. If the drying foliage touches the ground, it will acquire fungus spots that make the fiber, and the resulting paper, darker, duller, and spotted.

Allow the foliage to die back while it is attached to the bulbs. This method allows nutrients to nourish next year's foliage, while some of the elements that are undesirable in paper, such as starch, are removed. Cut the foliage at ground level when it looks sun-bleached.

6. Hand-beat the fiber on a hard surface for about 10 minutes or until done (see Is It Beaten to a Pulp? in chapter 3).

7. To form a sheet, use either Western or Eastern style (with formation aid). In addition to forming sheets with this fiber, you can color it, or add a bit or formation aid and use it as a pulp-painting medium (see chapter 6).

VEGETABLE/FRUIT PAPYRUS

Egyptian papyrus is made by overlapping pieces of the stalk of the papyrus plant and pounding or pressing the pieces together. A similar technique can be used to make vegetable papyrus from common household vegetables or corn husks.

These are not true papers, since the fibers are not beaten to a pulp and made into sheets of paper on a mesh surface. These sheets of paper are beautiful when held up to the light. Slices of vegetables or fruits are laminated together and when pressed become paper-thin, revealing the intricate patterns of the seeds and veins.

You Will Need:

Choice of vegetables or fruits (see box at left)
Sharp knife
Cutting board
Stove or hot plate
Pot of water
Vegetable steamer (optional)
Slotted spoon and/or strainer
Old felts or other couching material (such as interfacing)
Press
Interfacing (2 sheets per piece of papyrus)
Grater (optional)
Peeler (optional)

VEGETABLES AND FRUITS FOR PAPYRUS

When selecting fruits and vegetables for papyrus, choose those that are firm when raw but soften when cooked.

Apples   Peppers (red
Beets   and green)
Carrots   Squash
Onions   Star fruit
Oranges   Turnips
Parsnips   Zucchini

You can add as much texture and color to your vegetable papyrus as desired. Try these variations:

◆ Grate or peel the vegetables, or slice them lengthwise.
◆ Mix and match vegetables to make a multicolored sheet.

1. Remove carrot tops, apple stems, and other undesirable parts, as necessary. Slice the vegetables into ¼" rounds. (Some think that the vegetables are sliced extremely thin, but it is the pressing that makes them thin.)

2. Bring a pot of water to a boil, and drop the sliced vegetables into the water. Cook the rounds until they are soft to the prick of a fork. Softer vegetables (zucchini or squash) will need only a couple of minutes, but firmer ones (carrots or beets) may require up to 10 minutes. When done, remove the rounds from the water with a slotted spoon, or strain them.

3. Don't use good felts for this project — some vegetables (such as beets) stain. Arrange the vegetable rounds on your felt or other couching material, making sure they overlap by at least ¼". Place a second felt on top. Since the vegetables are still quite thick, either press the sheets individually, or stack several felts between the layers.

4. Press the sheets.

5. Remove the sheets from the press, and transfer them to dry sheets of interfacing. I recommend placing them between sheets of interfacing and drying them in a stack dryer or a press — vegetable papyrus tends to be sticky and difficult to remove from blotters or boards when dry. If you air-dry vegetable papyrus, it will shrink, resulting in wavy potato chip–like pieces.

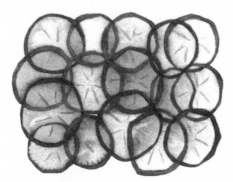

zucchini papyrus

TIP

Use a vegetable steamer to cook the vegetables, particularly if you want to keep track of certain pieces, since they will stay put in a steamer instead of floating around in a pot of water.

TAMALE PAPYRUS

Papermaker Andrea Peterson developed this technique for making a pseudo-papyrus that resembles the traditional papyrus in look.

You Will Need:

1   package tamale wraps (available in grocery stores)
      or corn husks
1½ gallons of water
1   cup soda ash
Bucket of clean water
Newspaper
Interfacing
Wheat paste
Small paintbrush

1. Trim the pointed tops off the tamales or corn husks.

2. Soak tamales or corn husks in water for a couple of hours.

3. Heat water in a nonreactive pot (stainless steel, glass, or enamel), and add soda ash just before boiling. Add the wet tamales/corn husks, and simmer for 30 minutes. This quick, harsh cooking will make the fiber translucent and soft.

4. Rinse the tamales/corn husks gently by dunking them in the bucket of clean water. They are fragile at this point, so handle them carefully.

5. Take a few pieces of tamales/corn husks, and pat them dry between sheets of newspaper.

6. Lay the tamales/corn husks on a dry piece of interfacing. Notice that one side of each piece has ridges, and the other is smooth. Place the side with the ridges toward the interfacing, so the smooth side is up. Lay five or six pieces next to each other so that they just touch.

7. Brush wheat paste onto the layer of tamales/corn husks. Place another layer perpendicular to the first, this time with the smooth side down (smooth side touching smooth side). Place a layer of newspaper and then another sheet of dry interfacing, continuing to build a stack of up to five sheets.

8. Place the tamale/corn husk papyrus in a press or under a weight (heavy books or a bucket of water), and keep it under pressure until dry. Place a piece of newspaper between each layer to wick away moisture as the sheets dry, and change newspapers daily. It will take a couple of days for the sheets to dry. You can also dry them between blotters in a drying system (see chapter 1).

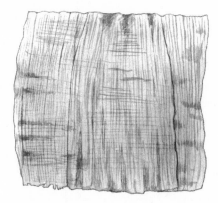

Quick cooking is the secret to making tamale or corn husk fibers soft and translucent.

CHAPTER 3

BEATING PULP

Beating the pulp is an essential step in the papermaking process. The fiber (cooked or not) needs to be turned into a loose viscous matter that can be scooped onto a screen to form sheets of paper. This can be achieved by several methods, including hand-beating; using simple household equipment, such as a blender; or using specialized equipment designed for papermaking such as a Whiz Mixer, a stamper, or a Hollander beater.

The equipment you use to beat the fiber — and the length of time — will vary the resulting paper. Particularly when you use a Hollander beater, lengthening the beating time can produce dramatic results. Abaca fiber beaten for 30 minutes will produce a soft, opaque sheet that shrinks very little; if it is beaten for 8 hours or more, the resulting sheets are crisp and translucent, with a high shrinkage rate.

A SHORT HISTORY OF BEATING

The first papers in China were beaten by hand with a mortar and pestle. Later, Chinese papermakers used trip hammers — long foot-powered beams on pivots — to stamp the pulp. In some areas, papermakers simply pounded the fiber on flat surfaces with mallets. In Japan, such arduous hand-beating led to the development of mechanically operated stampers.

Water-powered stamping mills were first used in Europe. The advent of printing from movable type in the 15th century created a large demand for paper, and the stampers were redesigned and improved to increase production.

Wind-powered mills in Holland lagged behind the water-powered mills of Germany. In an effort to keep up with the Germans, the Hollander beater was developed in the late 1600s. This machine greatly increased production and reduced the need to ferment cotton and linen rag fibers. Hollander beaters are still in use today at most large-scale papermaking studios. They take up a fair amount of space and are quite expensive.

In Japan, the *naginata* beater has rotating knives that neither shorten nor cut the fiber. It is used after stamping to tease bast fibers apart.

HOW BEATING WORKS

Individual papermaking fibers are slender, just millimeters in length. They are made up of cellulose molecules — long chains of hydrogen and oxygen atoms — with tiny fibrils resembling pipe cleaners, which interlock and bond when formed into sheets of paper. This bonding is called hydrogen bonding, and it occurs naturally between cellulose and water molecules because they have similar polar charges, which allow them to attach together like magnets. As the cellulose molecules are beaten, tiny fibrils on the fiber surfaces are raised, creating sites for water molecules to attach.

Beating also shortens the fibers, so they are evenly distributed during sheet forming. As water is removed and evaporates throughout pressing and drying, the fibers are actually pulled closer together, and hydrogen bonds form between the cellulose molecules.

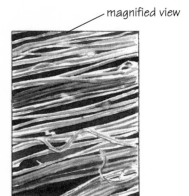

magnified view

In hydrogen bonding, hydrogen atoms are attracted to each other and, in the process of bonding, hold water and fiber molecules together.

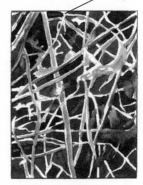

magnified view

As papermaking fibers are beaten, they become shorter and are consequently more evenly distributed as the sheet is forming.

A few fibers can be beaten right after they are obtained, without having to be cooked first:

- Recycled paper
- Prepared sheet pulp from a papermaking supplier (cotton linter, abaca, etc.)
- Some plant fibers, such as flax or hemp — but only if you beat them in a Hollander beater (otherwise they require cooking)
- Old rags (new rags should be cooked to remove starches, etc.), which also require beating in a Hollander

HAND-BEATING

Plant fibers can easily be beaten by hand. Although it is labor intensive, hand-beating draws out the fibers, leaving them long, strong, and tenacious. You can hand-beat your plant fibers using a mallet. Papermaking supply companies sell special mallets for hand-beating, but a baseball bat or a meat tenderizer will also work. Just be sure the wood won't crack or wear down, because the residue will contaminate the plant material. Work on a hard, clean surface — such as a hardwood cutting board or Formica — that will not split or splinter.

Before you start beating, set aside a bit of the plant material in a jar with a lid. After beating the remaining fiber for a while, you might think your efforts are fruitless. But when you compare some of your pounded pulp with your reserved sample, you'll be able to gauge your progress.

You Will Need:

Suitable fiber
Pounding surface (table or countertop)
Bat or long stick
1 or 2 mallets (optional)
Water
Large bucket

1. Take a ball of unbeaten pulp the size of a cantaloupe, and squeeze out as much water as possible. You can press the water out in a small press if you have one, or you can squeeze it by hand. The pulp should not splatter when you start beating. If this happens, your fiber is still too wet.

step 1

2. Spread the mass of pulp out onto a table or countertop, first making sure the surface is clean. The initial beating loosens the fiber strands and spreads them out. Pound the lump with a bat or a long stick. Use some force, but allow the weight of the stick to do most of the work. Starting at one end of the mass of pulp, move across it a stick's width at a time. When the fiber has spread out, fold the top, bottom, and sides up on themselves and start again. Beat for at least 15 minutes, until the plant material is thoroughly loosened.

step 2

3. At this point, you can switch to a hand mallet for more refined beating, but you can also continue with a stick or a bat. Some people use two mallets, one in each hand. Add water to the pulp during this stage of beating. Sprinkle about ½ cup of water (this will vary according to the type of plant fiber) into the center of the mass, fold it up, and turn it over. Begin pounding. Beat the fibers directly against the pounding surface, spreading and flattening them as you work. Add water as the pulp dries out, folding up the mass and turning it over again. The batch of pulp will begin to look homogenous, a mass of tiny fibers lumped together. Beat for 15 minutes, and perform the pulp test (see Is It Beaten to a Pulp?). Continue beating if necessary.

4. Your pulp is now ready for papermaking. Put the pulp in a large bucket, fill it with water, and stir it vigorously with a stick to loosen the fiber and get rid of any knots. Add the pulp to the vat (see different methods for filling the vat in sections on Western and Eastern papermaking).

MAKING PULP WITH A BLENDER

Blenders are limited in their ability to beat pulp. The action of a blender is to cut and chop, rather than pound and beat. Blenders are suitable for recycling paper, and even for preparing preprocessed sheet fibers, but a blender does not add strength the way beating does. Beating by hand, or in a Hollander beater, opens up more hydrogen bonding sites on each fiber, yielding a pulp that makes a stronger paper.

For blender beating, the fiber should be cut into small pieces. (This can be done before the fiber is cooked, when it's dry and easier to cut, or after you cook it, when it is wet.) Cutting prevents the fiber from tangling in the blades of the blender.

You Will Need:

    Suitable fiber
    Blender (one you won't use for food)
    Water
    Bucket or vat for collecting blended pulp
    Strainer
    Fine mesh (silk-screen mesh or mosquito netting,
        sold in camping stores)

1. Fill your blender three-quarters full with water, and add a handful of pulp. Break the pulp up a bit, so it's not one big lump. Put the lid on the blender, and turn it on. I usually start at a slow speed — such as mix — and then switch to a higher speed when I hear the blender running smoothly (after about 10 seconds). If the motor sounds strained, turn it off and check that the pulp is not wrapped around the blades, or that there is not too much pulp in the blender.

The blending time varies, depending on the fiber and how coarse you want your paper to look (for example, onion skins will break down to pulp in just a few seconds; iris leaves can be blended for a

It is best not to use your kitchen blender for beating pulp, since residues in the fiber could contaminate your food.

minute or more). Experiment with blending more and less to create finer and coarser paper.

2. You can dump the freshly beaten pulp directly into the vat if you will be making paper right away; or you can collect it in a bucket. Continue blending the pulp by the handful. If you will not be making paper right away, strain the pulp and store it in the refrigerator. Line your strainer with fine mesh to keep the tiny fibers from falling through the strainer holes.

SUITABLE FIBERS FOR BEATING

The type of fiber you use can determine the best beating method.

Hand-Beating
✦ Plant material (cooked and rinsed)

Blender Beating
✦ Recycled paper (torn or cut into 2" squares)
✦ Preprocessed fibers
✦ Plant material (cut into 1" lengths, cooked, and rinsed)

Hollander Beating
✦ Proprocessed fibers such as cotton linters and abaca

✦ Cotton or linen rag
✦ Raw, tough plant fibers, such as flax or hemp

Beating with Drill Attachments and Whiz Mixers
✦ Preprocessed fibers, such as cotton linters and abaca
✦ Prebeaten pulp (dried, refrigerated, or frozen)

Beating with Stampers
✦ Plant material (cooked and rinsed)

USING BEATING MACHINES

There are several mechanical beating aids. Some — such as the Hollander beater — are traditional; some are borrowed from other crafts or developed as inexpensive alternatives to the Hollander beater.

Hollander Beater

A Hollander beater is an oblong tub, rounded at both ends. A roll with flat-ended blades sits in the tub, directly above a bedplate. The fibers circulate in the water-filled tub and are macerated, shortened, and hydrated as they pass between the roll and bedplate.

You can create very different pulps — for example, opaque or translucent, crisp or limp — by varying the beating process (though some of this is determined by the fiber itself). The trick is in adjusting the beater roll so that it is closer to or farther from the bedplate to alter the bruising action. If you are beating a well-worn soft cotton rag, leave the roll up higher, so there is less cutting. For a very translucent abaca, you would need to set the beater roll close to the bedplate, then beat the fiber for several hours. The longer you beat the fiber, the more translucent it becomes, and the more it will shrink as it dries unless it is restrained (see Drying in chapter 5).

Drill Attachments and Whiz Mixers

Drill attachments used for mixing paint also work well for breaking down sheet pulps such as abaca and cotton linters. These are not suitable for preparing plant fibers or rags, however, because they do not beat the pulp — they simply rehydrate already-beaten pulp.

Whiz Mixers are made specifically for papermaking and are available from papermaking suppliers. Some papermakers have also used garbage disposals to recirculate and mix pulp.

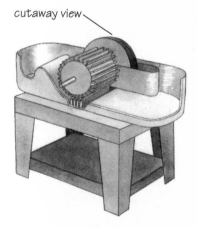

cutaway view

If you decide to set up a production facility, you should consider a Hollander beater to make your beating operation easier and faster.

When using a drill attachment or Whiz Mixer, just fill up a garbage can or bucket with water, add the fiber, stick the end of the attachment in the solution, and turn it on. Stop beating when the pulp is adequately loosened.

Stampers

Stampers provide a mechanized system for pounding fiber, and treat it much as if it were beaten by hand. Stampers are commonly used in Japanese papermaking, since the long bast fibers need relatively little beating to be separated. Stampers are available in this country from Harry Wold (see resource guide).

WHERE TO STASH THE BEATEN PULP

For best results, you should make paper right after you prepare the pulp. This is not always practical, however, and there are many methods of storing pulp. If it will be used within a few days, and the weather is not too hot and humid, you can just leave the pulp at room temperature. But if you leave it too long, it will start to ferment.

If you live in a hot and humid climate, or if you will not be using the pulp within a few days, it is best to strain it (if it's diluted in water) and store it in the refrigerator. It will keep this way for a couple of months. Line a strainer with a fine mesh to catch the small particles and prevent them from clogging your drain. Place the strainer in a bucket, or — if you have a drain — drill holes in the bottom sides of the bucket, and let the water flow onto the floor and into the drain. Strain the pulp, and store it in a plastic bag or bucket with a lid. Label the pulp, specifying the date and type of fiber used.

Some papermakers freeze pulp for longer storage. If frozen, it may need to be rebeaten to loosen the fibers. You can also dry your pulp by breaking it into small clumps so that it will dry thoroughly. In most cases, dried pulp will also need to be rehydrated and beaten.

TIP

If your stored pulp starts to smell, you can "freshen" it by adding a few drops of bleach, mixing well, and then thoroughly rinsing so that the bleach does not continue to break down the pulp.

CHAPTER 4

USING ADDITIVES

$\mathcal{S}$everal additives can be blended into your papers to color and coat them and protect them from deterioration. These additives are not necessary, but they will change the quality of your pulp and, in many cases, enhance your papers. Most of them can be blended into the pulp at the end of the beating cycle and before you start making sheets. If you wish to add something to only one particular batch of pulp, you can stir it directly into the vat.

SIZING

Sizing is a liquid substance that coats fibers and makes them more water-repellent and bleed-resistant. It allows you to use watercolors and inks on the finished sheets.

Internal Sizing

Internal sizing is mixed into the pulp, coating the wet fibers before they become sheets of paper. It can be added to pulp at the end of the beating cycle and before sheet formation. Synthetic internal sizings (alkylketene dimers) are available from papermaking suppliers (see resource guide).

If you hand-beat or process your pulp in a blender, disperse the entire batch of pulp in a bucket of water, and then add sizing according to the supplier's directions. Then dilute the pulp with more water in the vat for sheet formation.

If you use a machine to prepare pulp, you can add sizing according to the supplier's directions to the entire batch at the end of the beating cycle, letting the machine thoroughly mix it into the pulp.

Surface Sizing

Sizing can also be applied to papers after they are formed into sheets and dried. This technique, called surface sizing, adds an additional layer of protection against contaminants in the air. Traditionally used throughout Asia and Europe, surface sizing may have contributed to the permanence and durability of many historical papers.

This technique coats the paper and gives it a sturdier finish than internal sizing. Surface sizing can also be purchased from papermaking suppliers (see resource guide).

MAKING AND USING GELATIN SIZING

Here's how to make a standard 2½ percent gelatin solution:

You Will Need:
> Nonreactive cooking pot (stainless steel, enamel-coated, or
>   glass)
> 2   quarts cold water
> 7   packets gelatin
> Dry sheets of paper
> Sponge or wide brush

Put the water in the cooking pot; then dissolve the gelatin in the water, and allow it to swell for at least 10 minutes. Heat the solution to just under a boil, and keep it hot — if the gelatin cools, it will start to thicken, and application will be difficult.

WHAT TO USE?
Some common materials used for sizing include corn, rice, and wheat starch, as well as gelatin.

step 1

step 2

There are two techniques for applying gelatin to paper:

1. Work on a clean, nonporous surface. Wring out a sponge or wide brush in hot water, and dip it in the gelatin solution. Brush or wipe the gelatin onto the surface of a sheet of paper in even strokes, applying it in one direction. If you wish to coat both sides of the sheet, turn it over and brush gelatin on the other side as well.

2. Pour the gelatin solution into a shallow tray or vat that is larger than your sheets of paper. Place a sheet of paper in the solution, and turn it over to coat the other side. Leave it in the solution for 1 minute, then lift it out and allow the excess solution to drip back into the tray (otherwise, you will get a buildup of gelatin on the bottom edge). You can use a dowel or plastic tube to aid in lifting the sheet out of the solution.

After applying sizing with either method, stack the sheets one on top of the other, and press them to remove any excess gelatin. Hang your paper with clothespins on a line. Hang them loosely, with a slight bow at the top edge, because they will shrink as they dry.

BUFFERS, BRIGHTENERS, AND FILLERS

Some additives can make your paper smoother, more opaque, and resilient to environmental conditions that cause degradation. All these materials can be purchased from the papermaking suppliers listed in the resource guide, and they come with instructions for use. Add enough to treat the entire batch of pulp at one time.

Calcium carbonate ($CaCO_3$) comes in powder form and can be added to the pulp at the end of the beating cycle. It protects the paper from acidic contaminants in the air by slightly increasing the paper's alkalinity. It also acts as a filler, occupying the subtle crevices between the fibers and thus making sheets of paper smoother and more opaque. You can purchase calcium carbonate from papermaking suppliers or in ceramic supply stores.

Clay, or **kaolin,** also comes in powder form; it can be used to make paper opaque and smooth. It is also useful in paper casting, because it reduces shrinkage during drying.

Titanium dioxide (TiO_2) is a white pigment that can be added to make paper whiter, opaque, and smooth.

OTHER ADDITIVES

A number of other additives, each with its own specific purpose, can be used in the papermaking process.

FORMATION AID

This slimy substance is added to the vat during use of the Japanese sheet-forming technique, or any other techniques if you wish to slow the drainage time. This leaves more time for dispersing fibers during sheet forming. It also prevents the fiber from clumping. The traditional Japanese formation aid is *neri,* which is extracted from the roots of the Japanese *tororo* plant *(Abelmoscshus manihot Medikus* or *Hibiscus manihot L.)* by a process of pounding and soaking them in water. In a few hours the water becomes a thick, gooey slime called *tororo-aoi.*

A synthetic formation agent is available from papermaking suppliers in powder form (see resource guide). There are three types: PNS, for use with pulps that have no other additives, such as sizing or retention aid; PEO; and PMP. Both the latter can be used with pulps that do contain other additives.

There are also natural alternatives to *neri,* such as the fruit of the okra plant and the roots of hibiscus and hollyhock.

Synthetic Formation Aid

If you want to make your own formation aid, here's what you'll need:

You Will Need:
- 1 tablespoon powdered formation aid
- ½ gallon water
- Empty milk or juice carton

step 2

1. Fill an empty milk or juice carton with water, and sprinkle in the powdered formation aid. Refrigerate it and let it sit overnight. (You can also sprinkle the powder into a blender filled with water, blend the mixture, and then dilute it as necessary.)

2. Before using, strain the formation aid through a fine mesh bag (an old pillowcase will do the trick) to prevent any clumps or powder bits from getting into the vat.

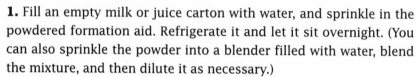

HOW MUCH FORMATION AID?

The amount of formation aid required depends on the type of fiber used and how it was prepared, as well as the thickness of the sheet and the quality of the paper you are making. More formation aid means the sheet will drain more slowly — which is useful when you are making thinner papers, because it allows time for even distribution of fibers. The obvious drawback is that if you use too much formation aid, it will not be possible to form a sheet.

If you use less formation aid, the sheet will drain faster — which is okay if you are making thicker sheets, because the sheet itself will slow drainage as pulp collects on the surface of the mould. However, since less formation aid means faster drainage, there is less time to cast off knots and lumps. You will need to replenish the formation aid in the vat every time you add pulp. Over time, you will develop a feel for how much formation aid to use.

RETENTION AID

Retention aid (also known as retention agent) is a cationic substance (made up of positively charged ions) that binds pigment to the fiber's surface. Pigments tend to have a negative charge and will therefore attach to the pulp if the proper amount of retention agent is added. Retention aid is available in liquid or powder form from papermaking suppliers. See page 78 for instructions on how to use it.

SODA ASH (NA$_2$CO$_3$)

This powder is a base substance that aids in the removal of noncellulose materials from a plant fiber when it is soaked or cooked in solution with the fiber. It is the most common alkali used in papermaking.

METHYL CELLULOSE

Methyl cellulose is an archival water-based adhesive that is reversible (you can "unglue" it). You can use it to size paper externally, strengthen bonding when casting with pulp, and attach paper to paper in the wet or dry state. It comes in powder form.

METHYL CELLULOSE GLUE

You can obtain methyl cellulose from any papermaking supplier or in some art supply stores.

You Will Need:
- 8 teaspoons methyl cellulose
- 1 cup boiling water
- 1 cup cold water

To make a glue, sprinkle the methyl cellulose into the boiling water; then add the cold water, and mix well. Let the solution cool to room temperature. It will thicken and become the consistency of honey.

MAKING PAPER HOLD UP TO HISTORY

Papers that are pH neutral (neither acidic nor alkaline) are considered archival, because they will not deteriorate over time. Noncellulose materials — acidic or alkaline — eventually cause the cellulose to break down, resulting in the paper's becoming discolored and brittle.

Papers made from wood (most commercial papers) are often treated with acidic or alkaline chemicals, which remain in the paper and break down the cellulose over time. In addition, some noncellulose materials are not removed from the wood, making the paper unstable and susceptible to decay. A typical example is a piece of newspaper. Leave it outside for a few weeks, and it will become crumbly and yellow.

Decay isn't a concern when paper is used for a day and then discarded. But it is a primary concern for artists and conservators who want their work to be preserved for hundreds of years. Acidity and alkalinity can be measured with a pH test strip, which is similar to litmus paper but measures pH on a scale from 1 (extremely acidic) to 14 (extremely alkaline). If you are concerned about the pH of your paper, you can test it at various stages of processing.

Factors to Consider

First off, you should check your water supply. Consult with your city water department or a company that sells water filters to determine how to treat your water. The pH of your water should be close to 7 (neutral). Ideally, it should also be free of pollution, microbes, particulate matter, chlorine, iron, and copper. Two of these contaminants will not even show up until long after your paper is made: iron causes foxing (small discolored spots) in paper and does not show up until the paper is dry, and chlorine breaks down cellulose over time.

Use pH test strips (Hydrion Short Range Paper is one commercially available brand) to test your cooking and washing solutions as well as your sheets of paper. During cooking, the pH of the cook solution should be between 10 and 11; below 10 is too weak, and above 11 could damage the cellulose. During washing, you can test the rinse water until the pH drops close or equal to the pH of your water supply. After making a sheet of paper and couching it (see chapter 5), lay an indicator strip on top of the sheet so that it absorbs some water, and then compare it with the Hydrion Chart. A pH between 7 and 8 is considered pH neutral for paper. A pH reading of 7 is truly neutral, but a slightly alkaline paper will be buffered against the acidic contaminants in the atmosphere. If the pH is over 8.5, the paper will deteriorate at a faster rate.

PIGMENTS AND DYES

There are several methods of coloring paper, but the two main tinting agents are pigments and dyes. Color permanence is of great concern to many papermakers. In general, pigments are more permanent than dyes, especially in terms of being lightfast (not fading over time); these are the most typical colorants used by hand papermakers.

A thorough reference guide to many coloring agents, including recipes, can be found in Elaine Koretsky's book *Color for the Hand Papermaker* (see reading list).

Pigments

Pigments are insoluble particles, which, when added to pulp, do not naturally attach themselves to the fiber. They must be physically attached to the fiber with a retention agent (available from paper-making suppliers; see resource guide).

Pigments are mixed with various dispersing agents for a variety of art mediums. They are mixed with linseed oil to make oil paint; ground with an acrylic emulsion to make acrylic paint; and ground with gum arabic or other binders to make inks and watercolors.

You can make your handmade papers colorful by adding pigments to them.

Pigments mixed with water are called aqueous dispersed pigments; these are the ones used for papermaking. Available from papermaking suppliers and some art stores, aqueous dispersed pigments are safe and easy to use.

AQUEOUS DISPERSED PIGMENTS

Aqueous dispersed pigments purchased from papermaking suppliers come with adequate instructions, but a few things are worth mentioning. You should add a little color at a time, and strain your pigment through a fine mesh (such as silk screen) to prevent any particles from getting into the pulp and forming blobs of color.

Once you have added the pigment, check the color of the water — it will probably have some color in it. A properly pigmented pulp has clear water, which means that all of the pigment has attached to the pulp. If the water is not clear (which will be true in most cases), you need to add retention aid. You do not need very much; start with a few drops.

Some papermaking suppliers suggest adding retention aid before adding the pigment, and others tell you to add it after pigmenting. I prefer getting the hue and intensity of the pulp to my satisfaction before adding retention aid, but I don't think there is a right or a wrong method. Some sizings give the pulp a positive charge and act as retention aids themselves, so you may not need to add additional retention aid if you use sizing.

DRY PIGMENTS

You can make your own aqueous dispersed pigments with dry powder pigments, which come in a much wider variety of colors. *Use caution:* Wear at least a dust mask and rubber gloves when working with dry pigments, many of which are toxic and should not be absorbed by the skin or inhaled.

When mixing dry pigments, disperse the pigment in enough water to make a thick paste (a few drops of water per tablespoon of dry pigment), and finely grind it with a mortar and pestle. When adding the ground mixture to pulp, strain it through a fine mesh (such as silk screen) to remove any particles that are still present.

Note that dry pigments have different physical properties. Some are hydrophobic (that is, they repel water) and so don't mix easily in water, so you need to add a wetting agent such as liquid soap, denatured alcohol, or gum arabic. In addition, some dry pigments do not mix thoroughly into the pulp, leaving particles of pigment standing on the paper surface. Your supplier should be able to provide you with details.

Dyes

Dyes are soluble substances that penetrate the fiber structure and chemically attach to the cellulose molecules, becoming part of the material. Most dyes are not lightfast. And most are toxic, so make sure you read the precautions and protect your skin and face when working with them.

Many types of synthetic dyes are used in the paper and textile industries; they are all described in *Color for the Hand Papermaker* by Elaine Koretsky.

DIRECT DYES AND FIBER-REACTIVE DYES

Direct dyes, sold in powder form, are available in supermarkets or art supply stores (RIT is one brand). They are easy to use and come in a wide range of colors.

Fiber-reactive dyes are similar to direct dyes and also come in a variety of colors. They react with the fiber and water; consequently, considerable rinsing is required to remove the dye from the water. Procion is one brand that can be obtained from art supply stores.

PEARLESCENT PIGMENTS

Pearlescent pigment powders, which add a sheen to paper, are easy to use. Unlike other dry pigments, they can be added directly to the paper pulp.

FIBER-REACTIVE FIBER DYEING

Fiber-reactive dyes are useful for producing deep shades, such as indigo, black, and forest green — particularly with Asian fibers. The following recipe was developed by Mina Takahashi, a papermaker in New York City, who specializes in Japanese papermaking. Divide the amounts of dye and salt in half, if you wish, to get paler shades.

You Will Need:

3 nonreactive pots
4 tablespoons Pro MX Reactive Dye (see resource guide)
Water
2 pounds noniodized salt
6 tablespoons soda ash
1 pound beaten pulp, drained
Lined utility sink or deep colander with a fine mesh

1. Prepare the chemicals in nonreactive pots: dissolve 4 tablespoons dye in 2 cups water at 95°F, dissolve 2 pounds plain noniodized salt in 2 gallons water at 95°F, and dissolve 6 tablespoons soda ash in 2 cups water at 95°F.

Much of the basic equipment for vat dyeing can be found in your home kitchen.

2. Add dissolved dye to salt water to create the dye bath.

3. Add the drained fiber to the dye bath, and stir continuously for 15 minutes. Add the dissolved soda ash in three portions, and stir for 2 minutes between additions. For maximum permanence and depth of shade, stir every 5 minutes for 1½ hours.

4. Drain the exhausted dye bath, and rinse the fiber well in a lined utility sink or colander.

step 3

NATURAL DYES

Many natural dyes can be created from items in your kitchen or garden — onion skins, walnut husks, or marigold, to name a few. In addition to plant substances, there are also animal substances that can be used for dyeing, such as cochineal and lac (produced from the bodies of insects). Although many natural dyes are not lightfast, they allow you to obtain hues not found in synthetic pigments and dyes. The deep brown created from walnut shells, the magenta resulting from cochineal, and the bluish gray from blackberry shoots are just a few examples.

Most natural dyes require the use of mordants (chemicals that fix the dye in or on a substance by combining with the dye to form an insoluble compound), which can react with the cellulose in the paper and cause deterioration. Elaine Koretsky has successfully used retention aid as a mordant with some natural dyes; her book *Color for the Hand Papermaker* (see resource guide) has detailed information.

BASIC DYE RECIPE USING NATURAL MATERIALS

Carol Tyroler, a papermaker in Washington, D.C., has experimented with many natural dye materials, including lichen, onion skins, and marigold petals, in recipes that use alum as a mordant. This recipe will dye 1 pound of fiber.

You Will Need:
- Nonreactive cooking pot (stainless steel, enamel-coated, or glass)
- 3 quarts water
- 1 pound dye-stuff (flower petals, onion skins, or other material of your choice)
- 4 ounces alum (available from hardware stores or suppliers — see resource guide)
- Colander

1. Put the water in the pot, and add the dye-stuff. Soak for about 30 minutes.

2. Add the alum, and bring the mixture to a boil, slowly. Cover and simmer for about an hour (the color may get darker if you cook it a bit longer; it will eventually reach a saturation point).

3. Remove from heat, and strain the dye-stuff from the dye bath, using a colander. Let the solution cool.

4. Once cool, the solution can be added to prepared pulp (mix it in well, to make sure the dye completely saturates the pulp), or you can dip dry sheets of paper into the dye.

If you are using Asian or natural plant fibers, cook the fiber with the dye bath. First cook it for 1 to 2 hours in soda ash (see chapter 2), until it is almost done. Then, rinse it thoroughly until the water runs clear, and cook it in the dye bath for another hour. Continue to soak the fiber in the dye bath for 24 hours; this will produce a stronger color. Rinse the fiber again until the water runs clear.

TRY THESE FOR DYEING

Queen Anne's lace

Zinnia petals

Lichen

Turmeric

Onion skins

Logwood

Carrot tops

Mint leaves

Red cabbage

Black-eyed Susans

Coreopsis

Parsley leaves

Ragweed

If you are using preprocessed fibers, such as cotton or abaca, soak the fiber in the prepared dye bath after you have strained out the dye-stuff. Let it soak 2 to 3 hours, and check the color. If you are satisfied, strain the fiber and rinse it until the water runs clear. If you want a deeper color, let it soak overnight in the dye bath.

You can reuse the dye bath by adding ½ ounce of alum per pound of fiber. With some materials, there will be enough color for several more batches of dyeing; with others, you will be able to dye only one more batch. The color may also vary with subsequent dippings.

Variations

◆ For marigold dye, use 1 pound fresh or dried marigold petals. After extracting the dye, you can add these petals to the vat for a decorative effect, if desired.

◆ For walnut hull dye, use 1 pound dry walnut hulls. Note that alum is not needed in this recipe, because the tannin in walnut hulls acts as a natural mordant.

VARIABLES THAT INFLUENCE COLOR

◆ Amount of time fiber is in dye bath
◆ Age of dye (fresh is best)
◆ Type of pot used in preparation (enamel-coated and stainless steel are neutral; other metals will alter the dye bath)
◆ Mordant used (mordants include alum, tannic acid, white vinegar, copper sulfate, calcium carbonate, and cream of tartar)
◆ Temperature of cooking solution
◆ Fiber being dyed
◆ Type of water (distilled or tap)
◆ pH of dye solution

INDIGO VAT DYEING

A range of blue colors can be achieved by dyeing with indigo. Since indigo is insoluble in water, it must first be reduced chemically to indigo white, or leucoindigo, in which form it is soluble in an alkaline solution. Fiber immersed in this yellow-green solution will be penetrated by the soluble indigo white, and when removed from the solution into the air, the indigo white oxidizes back to its insoluble blue form almost magically.

Barbara Bansley has used indigo in her work as a fiber artist, knitter, and papermaker for over 10 years and has maintained her indigo vat for almost that long. Simple maintenance keeps the vat ready when she is.

The dye vat works best at temperatures around 60 to 70ºF. The recipes below show how to prepare a 30-gallon vat (large) or a 3-gallon vat (mini). The procedures are identical — a stock solution and a dye vat are prepared separately, and then the stock solution is added to the dye vat — but the 30-gallon vat needs more time for settling.

It's essential to wear protective clothing including a face mask and work in a well-ventilated area when handling lye.

You Will Need:

   2   small 16-ounce plastic containers

   Plastic garbage can or pail with lid (30-gallon for large vat; 5-gallon for mini vat)

   Stainless steel or enamel-coated pot (2-gallon for large vat; 2-quart for mini vat)

   Glass measuring cup (1 or 2 cup)

   Stainless steel or plastic spoons for mixing chemicals

   Wooden pole long enough to hold and stir while the end touches the bottom of vat

   Eyeglasses or protective goggles

   Dust mask

   Rubber gloves (kitchen type)

   Beaten and strained wet paper pulp

   Indigo (synthetic indigo works best with this recipe)

   Lye (sodium hydroxide beads)

   Thiourea dioxide

   Denatured alcohol (available at hardware stores — might be needed to dissolve nonsynthetic indigo)

Preparing the Dye Vat

You must prepare the dye vat before you prepare the stock solution. **Caution:** Working with chemicals can be hazardous! Work outside or in a well-ventilated area, and wear a face mask, eye covering, and gloves when preparing solutions — the fumes from lye can be overpowering, and splashed lye can burn the eyes and skin. And remember: Always add lye to water, never add water to lye.

   Measure 6 cups of water into each of two small plastic containers. Using the measurements in the chart at right, add the lye to one container, and the thiourea dioxide to the other, and let them dissolve. Then pour the two solutions into your vat, which contains the rest of the water. The lye and thiourea dioxide solutions will create a reducing environment, eliminating residual oxygen in the water.

PREPARING THE INDIGO DYE VAT

LARGE VAT
Water: 20 gallons
Lye: 3/4 cup
Thiourea dioxide: 4 teaspoons

MINI VAT
Water: 3 gallons
Lye: 3 teaspoons
Thiourea dioxide: 1/2 teaspoon

PREPARING THE STOCK SOLUTION

LARGE VAT
Water: 6 cups
Indigo: 1 cup
Lye: 1 cup
Thiourea dioxide:
   2 tablespoons

MINI VAT
Water: 1½ cup
Indigo: 3 tablespoons
Lye: 5 teaspoons
Thiourea dioxide:
   ⅕ teaspoon

Preparing the Stock Solution

Using the measurements in the accompanying chart, mix the indigo, denatured alcohol, lye, thiourea dioxide, and water as follows:

1. In a stainless steel or enamel-coated pot, heat the water to 110°F — a little warmer than body temperature. Do not boil.

2. Place the indigo powder in the measuring cup, and add a small amount of water, working it into a smooth paste. (If you are using nonsynthetic indigo and it does not dissolve in water, try dissolving it in alcohol: 1 cup for the large vat recipe, and 1 tablespoon for the mini vat.)

3. Submerge the measuring cup and its contents into the pot on the stove, and gently dissolve the indigo paste in the surrounding water. Work slowly to ensure that no lumps will form. Stir gently to disperse completely. Remove the cup.

4. Pour ¼ cup of this indigo solution into the glass cup. Add the lye and the thiourea dioxide to the glass cup, and work it into a smooth paste.

5. Dissolve the lye and thiourea dioxide paste into the pot in same manner as in step 3.

6. Stir the solution gently for about 5 minutes, trying to avoid introducing air into the pot. The liquid will turn to a muddy green color with a metallic blue scum on the surface.

step 6

7. Remove the pot from the heat, and let stand 3 to 5 hours (3 for the mini vat, 5 for the large one). The stock solution should appear greenish yellow under the oxidized blue indigo on the surface.

8. Finally, add the stock solution to the dye vat in the vat or bucket you used. Use some of the water from the dye vat to rinse out any indigo dye solution that has settled to the bottom of the pot. Stir, and allow to stand 5 to 7 hours (3 to 4 hours for the mini vat).

REDUCING AGENTS AND ALKALIS

Several reducing agents and alkalis can be used for dyeing. Here are some chemical alkalis (with their household names in parentheses); consult the dyeing books in the reading list for recipes on how to use them. All chemicals can be obtained from chemical supply houses (see resource list).

Common Reducing Agents
+ Thiourea dioxide
+ Sodium hydrosulfite
+ Rit Color Remover, which contains sodium hydrosulfite (available at grocery and drug stores)
+ Zinc metal dust (I do not recommend using this, because it's highly flammable when in contact with moisture — most suppliers won't even ship it, and it's hazardous to water supplies)

Common Alkalis
+ Potassium hydroxide or sodium hydroxide (lye)
+ Sodium carbonate (soda ash), available from papermaking suppliers
+ Calcium hydroxide (slaked lime), available at building supply stores
+ Hydrated sodium carbonate (washing soda), available in grocery stores

DYEING FIBER

You are now ready to use your indigo vat. Try dipping your wet fiber or dry sheets of paper into the vat one or more times to create various shades of indigo blue. Wear rubber gloves when dyeing — indigo will stain hands.

1. Using a large wooden spoon, gently remove the froth from the top of your dye vat (the combined stock solution and dye vat will now be referred to as the dye vat). Place the froth in a small pan to return to the vat after dying.

2. Immerse the beaten and strained (wet) fiber into the vat for 1 or 2 minutes. Stir it gently, using a wooden pole. (If you are dyeing sheets of paper, see next page).

3. Lift the fiber out of the vat using the wooden pole, and hold it against the side of the vat, allowing the indigo solution to drain back into the vat without excessive splashing. Splashing will introduce oxygen into the vat, and you want to avoid this.

4. After draining the fiber, lift it out of the vat, and lay it on a surface that you don't mind staining — or hang it to oxidize for about 30 minutes. If you desire a darker shade of blue, dip the fiber again, repeating steps 1 through 4, after the fiber has turned blue. You can dip multiple times — in Japan, some cloths are dipped 18 times. If you desire a lighter shade of blue, reduce the subsequent dipping time to 30 seconds.

5. When the desired shade is reached, rinse the fiber in clear, cold water until it runs clear.

DYEING SHEETS OF PAPER

Barbara has experimented with dyeing sheets of handmade paper as well. You must use a durable pulp (with at least some percentage of fiber such as flax, hemp, cotton rag, or abaca) that will withstand the multiple dipping process. Wear rubber gloves when dyeing — indigo will stain hands.

1. Put the indigo solution in a tray large enough to accommodate your paper.
2. Submerge a sheet in the vat, and hang it to dry.
3. Repeat to get the desired hue.

Several dyeing techniques can be used to create decorative effects, such as tie-dyeing *(shibori-zome)* and fold-and-dip dyeing *(itajime-zome)*. Consult the reading list for ideas and instructions.

MAINTAINING YOUR VAT

Barbara has kept the same indigo vat in use for over eight years by adding alkali, reducing agents, and indigo when needed. The vat does not spoil; it simply becomes depleted. She has tightly covered her vat for half a year and ignored it. After testing the pH and adding the proper chemicals, she found the vat to be in good shape. Here's how to maintain yours.

1. After completing a dyeing session, return the skimmed-off froth and bubbles (the indigo "flower") to the vat.
2. Use a pH litmus tester (available at garden stores) to test the pH level of the vat — it should be at 10. If the litmus paper test shows the pH level to be below 10, add more lye; if the pH level goes above

10, add thiourea dioxide. Add each item in small increments, and let the vat sit for an hour before testing the pH again.

3. If a large amount of dyeing was done, the vat will be light blue, and the indigo powder, thiourea dioxide, and lye may need to be replenished. Maintain the proportions of 5:10:3 — indigo:lye:thiourea dioxide — by weight. For example, mix 1 tablespoon of indigo to 2 tablespoons of lye and 1⅓ tablespoons of thiourea dioxide with enough water to make a slurry.

4. After making any chemical additions to "sharpen" the vat, take the wooden pole and give the vat a healthy stir for about 30 seconds, creating a whirlpool in the middle of the vat and stirring up and dispersing any sediment that has collected at the bottom of the vat. Stir vigorously. When the froth on the surface has collected in the center of the whirlpool, stop stirring with the wooden pole and hold it steady in position about a quarter of the way from the side of the vat. This will force the froth to make a fluffy, new "flower" in the center of the vat. Cover the dye vat, and allow it to rest for at least 8 hours (4 hours for the mini vat) before reusing.

CHAPTER 5

MAKING PAPER

There are three basic types of papermaking: Western, Eastern, and deckle box. This chapter, in addition to explaining how to make paper using different types of moulds and deckles, will also familiarize you with more papermaking terminology.

Eastern papermaking (I'll refer to this technique as Japanese papermaking, which is a significant type of Eastern papermaking) and the deckle box technique (sometimes referred to as Tibetan or Nepalese papermaking) came before Western papermaking and were originally used with bast fibers, such as kozo, mitsumata, daphne, gampi, and hemp. These fibers produced smooth surfaces for the ink and brushwork common in Eastern countries. When papermaking found its way to Europe, where the stylus was the common writing tool, cotton and linen rags were the usual raw materials. These types of paper were also used when the printing press was invented.

Although each technique was developed for use with certain fibers, they are somewhat interchangeable. Western papermaking is suitable for many fibers, but when couching (transferring paper from a mould to a felt) becomes difficult, you might try Japanese paper-making, which is also ideal for making very strong, thin papers. The deckle box technique can be used to make very large sheets of paper, because you can pour rather than dip the sheets, thereby eliminating the need to lift a large mould and deckle. It is also good for making thick sheets, combining pulps, and making a sheet without a vat.

COUCHING

Couching is a fluid rolling action. Lay the sheet down from one side, press briefly, then roll the mould up and remove it from the felt. Once a cushion of sheets is formed, you can press during the roll and just lift the mould right up.

The Western style of papermaking was developed in Europe and is currently the most common technique used by professional paper-makers in the United States. Cotton and recycled fibers are commonly used in Western papermaking, and you can form sheets with many plant fibers using this technique. For instructions on how to make a Western-style mould and deckle, see chapter 1.

You Will Need:

Western mould and deckle
Vat
Prepared pulp
Couching material
Strainer (optional)

Filling the Vat

1. Fill the vat with pulp and water. The ratio of pulp to water varies, depending on how thick you want your sheets to be. You will always need a certain amount of water mixed in with the pulp to be able to go through the motions of forming sheets, and you can't make sheets that are thicker than the height of your deckle. The limits will become apparent as you experiment. (For techniques for making thin or thick sheets, see page 97).

2. Start out with a thin pulp, and make a test sheet to see if you are happy with the thickness. Then add pulp or water as neces-sary. It is easier to start thin and add pulp than to remove it. As you make sheets of paper, you will need to replenish the vat with pulp. I have my pulp in a bucket within reach of my vat. Sometimes you will need to strain out some of the water, so you are not diluting the solution in the vat as you add pulp. Keep a clean strainer next to your vat, so if this happens, strain some water out of the pulp before you add it to the vat.

step 1

step 2

TIP

Be careful when handling the pulp. It is in its freshest and best state right after beating. Avoid creating knots and clumps, which will end up as impurities in your sheets of paper. Keep the pulp suspended in water, and take care to keep bits of dried pulp from collecting on the sides of buckets or vats by continually spraying them down with a hose.

If you are going to mix different pulps together, keep in mind that you should always prepare the different pulps separately, because they might require different cooking formulas or beating times. For example, you might cook your kozo for 2 hours and then beat it; but daylily will need only an hour to cook and can be processed in a blender. Mix the different fibers or additives together just before forming sheets of paper. (See box on page 98.)

Agitating the Pulp

1. Pulp tends to settle to the bottom of the vat as it sits, so it needs to be agitated to evenly disperse it before you pull a sheet. This is called hogging the vat. To do this manually, stick your hand into the slurry of pulp and water, and vigorously mix it. The vat should be hogged each time you make a sheet, especially when you replenish the vat with pulp.

Preparing the Mould and Deckle

1. Before you make your first sheet of paper, thoroughly wet the mould and deckle, which will encourage even distribution of the pulp. (After you make your first sheet, the mould and deckle will be wet, so you do not need to repeat this process. Occasionally, however, you should spray your tools to remove bits of pulp that tend to collect.)

Agitating the pulp

Preparing the mould and deckle

2. Hold the mould with the screen side facing up, and place the deckle on top, making sure that all edges and corners line up. (A common beginner's mistake is to hold the mould upside down, which makes it impossible to couch.)

Making a Sheet of Paper

There are many ways to make a sheet of paper; this is my technique for making Western-style paper (the following sequence should take place in one fluid motion):

1. Dip the mould and deckle into the vat at a 45° angle to the bottom of the vat, scooping underneath the surface of the pulp and pulling toward yourself.

2. Bring the mould and deckle parallel to the bottom of the vat, lift it up out of the mixture, and shake — left and right and back to front — to interlock the fibers. The shaking motions are short, semivigorous shakes, as if you were a prospector panning for gold. Every pulp will drain differently, and the amount of time to shake will vary.

3. Shake until you see the fibers starting to settle on the screen as the water drains, but do not shake too long. If you continue to shake once the fibers have started to settle, you will damage the sheet. (See Kissing Off if you wish to return the sheet to the vat). The sheet of paper is being formed as the pulp settles on the screen and the water drains through it. Shaking interlocks the fibers and distributes them evenly on the mould, making the sheet of paper uniform in thickness.

4. When you see that the fibers have settled on the screen, balance the mould and deckle on the corner of the vat, or on another flat surface, and remove the deckle. Take care not to drip water from the deckle onto the freshly made sheet (the resulting marks are referred to as papermaker's tears). This can be accomplished by gently whisking the deckle over the mould while keeping the two parallel. If there is fiber on the edges of the mould, gently brush it back into the vat with your fingers. This helps keep excess pulp off the felts. Set the deckle to the side, or drop it into the vat to rinse it off.

step 1

step 2

step 4

KISSING OFF

If you are not satisfied with the sheet of paper that comes out of the vat, don't just dunk your entire mould and deckle back into the vat. Remove the deckle, flip the mould over, and smack the mould on the surface of the water. The sheet will fall into the water, and the mould will be clean. Be sure to stir the vat again to break up that sheet before pulling another one.

5. Gently tilt the mould, and watch to see that the sheet on the surface does not start to slip. If it does, return it to the horizontal position and continue to drain, or kiss off the sheet and begin again.
6. When you are able to tilt it, hold it at an angle with one corner pointing into the vat, letting a stream of water drip from the corner. When the stream slows to a trickle, the sheet is ready to couch.

step 5

step 6

ANOTHER WAY TO REST THE MOULD

An alternative method for resting the mould when removing the deckle is to place a thin strip of wood spanning the two sides at the back of the vat. The mould can rest between this strip of wood and the front edge of the vat. Some papermakers have a second "drain" vat, with one or two sticks spanning its width for resting the mould while they get the felt set up or form another sheet.

Couching

1. Dampen a felt, and smooth it out in the couching tray. The dampness aids in releasing the sheet of paper from the mould. As there is a significant amount of moisture built up in the pile, called the post, subsequent felts may not need to be dampened.

2. Set the driest edge (the top edge as you were draining) of the mould down on one edge of the felt, with the wet sheet facing the felt.

3. Holding the mould in the center of both long sides, lay the mould down so that the entire mould is flat on the felt.

4. Apply even pressure to the back of the mould's edges and screen to ensure that the sheet transfers to the felt. You should see a little water come up through the back of the mould.

5. Lift the edge of the mould that you set down first, peeking underneath to see that the sheet has released, and remove the mould, taking care not to drip on the newly formed sheet.

6. To continue making sheets, place another felt directly on top of the sheet you just made. Make another sheet, and then couch it onto the second felt, lining it up with the sheet underneath. You can create a guide for lining up the sheets of paper by using two pieces of tape to mark where to set the mould each time. Keeping your post stacked evenly is important; the sheets will couch and press better if they lie directly one on top of the other. Continue making sheets to form a post of paper.

As you make sheets, you are removing pulp from the vat. Remember to add more pulp to replenish the vat after every three or four sheets, in order to maintain consistency in the thickness of the sheets. The amount of pulp you add will depend on the size of the vat and the thickness of the sheets of paper. Each time you add pulp, stir the vat thoroughly to mix in the new fiber.

THIN SHEETS VS. THICK SHEETS

Couching can be frustrating, particularly with thin sheets of paper. If you are having difficulties, try this: Cut a piece of lightweight interfacing to the size of your mould, and set it on the mould as you would an adapted su (see page 18). Form a sheet of paper, and let it drain. Then, rather than couching it, simply remove the interfacing from the screen with the sheet attached to it, and set it on your stack of paper (hold it by 2 diagonal corners to lift). Or, couch both the sheet of paper and the interfacing. Don't bother removing the sheet from the interfacing until the sheet of paper has been pressed. You could even hang the interfacing with the sheet attached to dry, or slip the unit into your drying system as is, and peel the sheet off when dry (see drying techniques, page 107). For this technique, you will need to have several sheets of interfacing cut to the size of your mould.

To make thick sheets of paper, you can couch two sheets of freshly made paper together. When they are pressed and dried, the sheets will bond and become one. To make sure the edges line up and the sheet is uniform, create a couching guide (see page 124). Another way to make thick sheets is to pour the pulp directly onto the mould, thus "casting" your sheet rather than forming it. You can also make thick sheets by using a deckle box (see page 101).

Many plant fibers are well suited to the Japanese papermaking technique, called *nagashi-zuki,* which has traditionally used long bast fibers, such as kozo and gampi. Some long fibers are unsuitable for the single-dip Western method of sheet formation because they tend to tangle and form knots in the paper. If you experience this problem, try the Japanese method, which involves multiple dippings to build up a sheet. You'll also find that the Japanese couching method works better with some fibers. For complete information on traditional and modern Japanese papermaking, please refer to *Japanese Papermaking* by Timothy Barrett (see reading list). For instructions on how to make a Japanese-style *sugeta,* or how to convert your Western mould into a *sugeta,* see chapter 1, page 18.

You Will Need:
- Sugeta
- Vat
- Couching material
- Prepared pulp
- Stick or other stirring tool
- Formation aid

step 1

1. Fill a utility sink half full of water, and add 1 gallon of hydrated pulp. Using a stick or other stirring device, mix the pulp and water well for 2 to 3 minutes. Add 2 cups of concentrated formation aid (see page 74), and continue stirring vigorously. This will break up clumps of fiber. The slurry should feel viscous but not slimy. When you make your first sheet of paper, pay attention to the drainage, and add more

---

MIXING DIFFERENT PULPS

You are not limited to filling your vat with just one pulp. You can add some abaca or rag to a recycled or cotton linter pulp for more strength; add flower petals or bits of thread as a decorative item; mix two different pulps, such as daylilies and kozo, to create a unique sheet — the sky's the limit!

formation aid if necessary. If the sheet does not stay adhered to the mould surface while you are dipping, it may mean you need more formation aid.

2. To make a sheet of paper using a *sugeta* and the Japanese sheet-forming technique, place the flexible *su* on top of the *keta,* then close the *keta* so that the *su* is held tautly in place. With an adapted Western mould, place the flexible screen on top of the wire mesh mould, and secure it in place by sandwiching it between the mould and deckle.

step 2

3. The Japanese method involves dipping the mould into the vat repeatedly, building up layers of fiber. This is the essential difference between Japanese and Western sheet formation. The multiple laminations create more surfaces for fibers to bond, which is what makes these papers very strong.

The first dip is quick and is referred to as the "first skin" on the sheet of paper. To start, dip the *sugeta* into the vat as if you were forming a Western sheet, but when you pull it out of the water, tilt it away from you and toss off the excess water. You should see a thin film of fiber on the screen. This first layer slows drainage of subsequent dips.

step 3

4. Dip a second time, and hold the pulp in the deckle while rocking it back and forth to form the second layer. After a few rocks back and forth, toss the water off of the back edge again. Dip again, and this time shake left to right, then toss off the excess pulp. Alternating the shakes from back and forth to left and right will form a stronger sheet with less noticeable grain direction. If you wish, you can align all the fibers; to do so, you should shake only front to back.

Repeat this dipping and sloshing several times until your paper becomes the desired thickness.

step 5

5. Couching this type of paper is somewhat different from the Western couching method. Remove the deckle, lift the flexible screen at the taped or sewn edge, and gently roll it down onto a damp piece of nonwoven interfacing. Brush the back of the *su* with even strokes to ensure that the sheet of paper sticks to the interfacing. Apply a bit of pressure along the front edge of the screen to make sure it adheres to the interfacing, then peel the screen away from the sheet of paper.

Separate the sheets with felts or interfacing to build up the post.

Removing the screen by peeling it away from the paper works well for making very thin papers in both Eastern and Western styles.

TRADITIONAL JAPANESE-STYLE COUCHING

Traditionally, Japanese papers are couched one on top of each other, with just a piece of thread laid down on one edge to enable the papermaker to separate them after pressing. An entire post is formed with no felts. It stands and drains overnight and then is slowly pressed, with a gradual increase in pressure. After pressing, the sheets are peeled from the post and brushed onto wooden boards for drying.

USING A DECKLE BOX

If you only have a little bit of pulp (particularly with plant fibers or batches of specialty pulps), you can make sheets of paper using a deckle box. This is also a handy technique for crafting very thick sheets of paper, creating multifibered sheets, or making sheets that are too large to handle by dipping into a vat. For instructions on how to make a deckle box, see chapter 1.

You Will Need:
> Deckle box
> Prepared pulp
> Vat
> Couching material

step 1

step 2

1. Place the deckle box in a vat, and fill the vat with water so that it comes up to approximately 2" higher than the surface of the mould. The water will fill the deckle box as well.

2. Pour a sheet's worth of pulp into the deckle box. You can distribute the pulp over the screen surface as you pour it into the deckle box, or you can stick your hand into the box after you have added the pulp, and move it around (similar to stirring pulp in the vat).

3. Lift the box up, and shake as you would to form a Western sheet. If you discover that there is too little or too much pulp, simply resubmerge the entire box, add or remove pulp, and form the sheet again.

4. Remove the deckle, and couch as described in steps 1 to 3 on page 96.

Lining the mould and deckle with a plastic sheet, which you later pull out from underneath the fiber-water mixture, eliminates the need for a vat.

Another Deckle Box Technique

If you do not have a vat to set your deckle box in, you can line the mould and deckle with plastic sheeting to block the flow of water through the screen, simulating a deckle box. You can then pour a solution of fiber and water (with formation aid if necessary) onto the plastic. Then, simply whisk the plastic sheeting out from underneath the fiber (imagine the magician pulling the tablecloth out from under the table setting) in one fluid motion. As the fiber and water start to settle and drain, gently rock the deckle box back and forth, or shake it. Remove the deckle, and couch as usual. You can use this technique to form any size sheet of paper, but one good use is to form very large sheets of paper that could never be dipped in the traditional Western or Japanese fashion (see Making Paper with the Deckle Box, page 22).

PRESSING

Now that you have produced a post of paper, you need to press and dry the sheets. Pressing removes water from the sheets of paper and compacts the fibers tightly together as they bond. After pressing, the sheets will still be damp and will need to be dried.

All kinds of systems have been rigged for pressing, ranging from heavy books or rocks piled onto the post, to a bucket of water, to a hydraulic jack. The first presses in Europe were wooden screw presses, which took the strength of many men to tighten. Similar presses were used in the Far East, and in some countries, papermakers developed systems of drying sheets of paper directly on the moulds, eliminating the need for a press. Other systems involved drying sheets on boards or walls.

Some of the techniques described here are centuries old. Others are recent innovations, developed by contemporary papermakers who have designed systems to meet their needs. Some papermaking suppliers sell presses; for instruction on building your own simple press, see chapter 1.

Sponge Press

A low-tech device for pressing is a sponge. Use a clean sponge to avoid contaminating your paper. I recommend that you couch onto nonwoven interfacing when pressing with a sponge, because it is less absorbent and will require less sponging than a thick felt.

Take one sheet of paper with its interfacing, and place it onto a dry felt or newspaper (something that will absorb some of the water). Then, place a dry sheet of interfacing on top of the sheet of paper to protect the paper's surface, and gently but firmly sponge water out of the sheet. Repeatedly wring out the sponge, and continue pressing the rest of the sheet. Be careful not to damage the sheet by pressing too hard, but remove as much water from the sheet as you can. It is also important to press the sheet evenly, so that the whole sheet will dry at the same rate. The sheet is now ready to be dried. Repeat to press the remaining sheets, and then use any of the drying options starting on page 107.

Instead of a sponge, you can use a rolling pin or an ink brayer (available at art supply stores) to remove moisture from your sheets. Since these tools do not suck up moisture like the sponge, you will need to place the sheets between absorbent materials, such as newspaper, towels or felts, so that the moisture is transferred to the absorbent material as it is rolled out. Press gently at first; if you press too hard you will distort your sheet. To remove the most water from your sheet, change your blotting material a few times, and repeat the rolling out.

Using a Hydraulic Press

When using any type of hydraulic press, load the post of paper into the open press, then bring the press to full pressure. In some cases — for example, if your paper is very thick — you might need to press the paper gradually. This is very important, because if you

A sponge is a low-tech but effective device for pressing water out of your paper.

An ink brayer or a rolling pin can be used for pressing, but you will need to use absorbent material between the sheets to soak up the water.

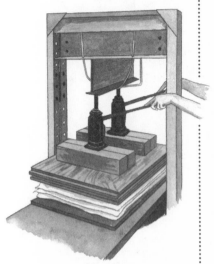

The advantage of a hydraulic press is that it can be used with an entire post of paper.

Be sure to get a good grip on the paper when peeling it from the felt so you don't damage the corner you are holding.

press too quickly, the water will explode out and damage the wet paper. When the press has been brought to full pressure, you can leave it for 15 minutes to an hour, depending on your press and the particular post of paper. You may not need to let it sit at all; just make sure that water is removed from the entire sheet. With most presses, the pressure hits the center of the sheet first. Allowing it to sit gives the water time to move from the center and seep out of the edges. Release the tension, and remove the post when you are done. If the sheets still seem very wet, you can press them again.

If you are pressing different types or sizes of papers, you might put a barrier of wood, metal, or firm plastic between the layers, so that they are in effect pressed separately. For example, if you are pressing plain white sheets together with pigmented sheets of paper, the pigmented sheets should be separated by a barrier so that the color does not bleed into the white sheets. In this particular case, the white sheets should be on top of the barrier to ensure that no color drips down onto them.

Although it is not necessary, fitting a hydraulic press with a gauge that measures the pressure can be handy, enabling you to record the amount of pressure used for different techniques. In general, you can press papers to full pressure, but if you are board- or spur-drying (see pages 110–11), you will need to adjust the amount of pressure applied.

HANDLING DAMP PAPER

After pressing, your paper is much dryer than when you first couched it, but it is still damp and very fragile and needs to be handled with care. When working with small or thin sheets, remove a sheet from a felt by starting at one corner on a short side of the sheet. Taking care not to damage the corner, gently get your fingers underneath it, and start peeling the sheet off the felt. I like to put at least three fingers behind the sheet and my thumb in front so that

I have a good grip. Hold the sheet taut, but do not stretch it.

Lift toward the other corner of the short side of the sheet, and when you get to it, take hold of it with your other hand. Continue to pull the sheet off the felt. To lay the sheet onto a surface, gently place the bottom edge on the surface, and roll the sheet down onto it, trying to avoid any creasing or bubbling. You can gently lift and reposition the sheet if necessary.

For thicker larger sheets, the traditional Western method of lifting sheets from a post is to grab two diagonal corners and gently lift the sheets. If your sheets are very thin and fragile, you can use a dowel or plastic tube to aid in lifting the sheet.

Place the tube along one edge of the sheet, and gently roll the edge of the sheet onto the tube. As you lift the tube, the sheet of paper will be lifted as well. If you couch on nonwoven interfacing, you can lift a sheet of interfacing with a sheet of paper on it in a similar fashion. You won't need to touch the corners of the sheet, and it should stay attached to the interfacing while you are moving it.

Lay the sheet of paper face down on the drying surface, and then carefully remove the interfacing — or put the blotter on top of the damp paper, and flip it over and remove the interfacing. Or, if it seems difficult to remove, just leave the interfacing attached to the paper — it will just take a little longer to dry. If you are drying your sheets against a surface, you can brush or roll it with a brayer while the interfacing is still on top, and then remove it (see drying technique on page 111).

As soon as you can, lift the other corner of the short side of the paper and continue pulling gently.

If you are making thick, large sheets, you can peel each sheet by grabbing two diagonal corners and lifting the sheet gently.

For thin, fragile sheets, use a dowel or plastic tube. Gently roll one edge of the sheet onto the dowel and lift both of them at the same time.

Felt Maintenance

It is very important to keep your felts clean because any residue left on them could end up in your next batch of paper. It is equally important to thoroughly dry your felts between uses so that they do not mold or mildew (it is extremely difficult to remove mold or mildew once it has started to grow). You can w͟   ͟ur felts in a washing machine, but use very little detergent anᴗ ͟lute it before putting it into the machine. Only use detergent if the felts have been stained or have a residue you wish to remove. Because machine washing tends to wear felts down somewhat quickly, you should check them periodically for wear and tear.

Many papermakers brush their felts with scrub brushes to get rid of pulp and paper fragments. This can eliminate the need to wash, but it too increases wear and tear. The gentlest cleaning method is to soak the felts in water and/or spray them with a hose.

To dry the felts, simply hang them between uses. A clothesline works fine, or you can build a simple rack system that is space efficient with a couple of strips of wood, some screws, and clothespins. This rack can double for loft drying paper, too (see page 110).

You can clean your felts by brushing with a scrub brush.

You can easily make a drying rack with two pieces of wood and some clothespins.

DRYING

There are many drying methods, and most of them are fairly simple. I recommend starting simply. There is no need to invest in an elaborate system — I have seen beautiful, high-quality papers that were made with basic equipment.

Some things to consider when choosing a system are climate, space, and the appearance of your finished sheets. The climate can affect how quickly your paper dries. The more humid or damp it is, the longer it will take to dry papers, unless you control the environment with an air conditioner or a dehumidifier. Where you will dry your papers is another consideration. You can hang them on a clothesline, dry them in spurs (four to five sheets pressed together), lay them out to dry on a table or a rack, brush them onto boards or walls, or set up a drying system (see below). The method of drying you choose will affect the texture of your papers as well. Experiment with different techniques to see which you like. Take several sheets of paper and try drying each one with a different method, and compare the results, noting differences in size, shape, surface, and texture.

Drying on the Mould

In some countries in the Far East, papers are not pressed at all but are dried directly on the moulds on which they were formed. The sheets stick to the screen surface as they dry. When dry, the sheets can be peeled off the moulds. This system requires many moulds, and it is typically used in arid climates where the sun dries the sheets rapidly, so that the moulds can be reused. Although the system requires the   e of many moulds, it does eliminate the need for a press and a drying setup.

If you live in a hot, dry climate, you can leave your paper on the moulds to dry.

DRYING ON THE MOULD

Papermaker Denise DeMarie of Newport, Oregon, produces a line of papers using primarily grass fibers, which she dries on the mould. This method allows surface textures to remain intact, and any three-dimensional objects (twigs, flowers, stalks) added to her papers are not disturbed by pressing. It yields particularly beautiful effects when long, wispy fibers and embellishments are used.

DeMarie recommends setting up a drain vat next to the papermaking vat and laying two sticks across the top, on which you can set the mould. Make a sheet of paper, and quickly set it on top of the sticks over the empty drain vat. Run your fingers around the edges of the deckle to tidy up any overhanging fibers. Do this quickly before the water drains out of the paper, to avoid pulling extra fibers away from the interior of the sheet. Once the water has drained, remove the deckle and set the mould in the sun — on end if possible — to dry.

Rotate the mould after a couple of hours to ensure even drying. Once the paper has thoroughly dried, gently peel it off the mould. Run a flat knife around the edges to start the peeling process.

DeMarie has developed a simple system for producing multiple sheets. Cut pieces of mesh (plastic needlepoint canvas or aluminum window screen) to the outer dimension of the mould. Sandwich a piece of this screen between the mould and deckle, and form a sheet as described above (similar to the adapted su technique described on page 18). After removing the deckle, lift the extra screen from two diagonal corners, and lay it on a large rack (or on any surface that allows air to move around the sheet). This technique is most successful for making small sheets. It is important to square up the wet sheet and screen every time you move it. If the paper and screen roll up while drying, peel the sheet off the screen, gently mist the sheet to relax it and flatten it between blotters under a weight.

Vacuum Drying

Another process I've seen that did not require a press or a drying setup was developed by Nance O'Banion. She had a series of 4' x 8' wooden frames with silk screen stretched across them. She couched freshly formed sheets of paper directly onto the silk screen, which could be propped against the a wall. Many sheets could be couched onto a single screen. Then, either they were left in the sun to dry, or the water was sucked out of them from the back by running a wet-vacuum cleaner over the screen.

Air-Drying

Another simple method of drying involves letting the sheets air-dry on the surface upon which they were couched. I recommend couching onto interfacing if you are going to air-dry, because it dries quickly. You can spread the sheets out on screens or on a rack so that air circulates around them, or you can hang them on a clothesline to dry. You can also pin the interfacing to a wall or board, to hold it taut as it dries. When the sheets are dry, just peel them off the interfacing. The sheets might cockle (curl) slightly — you can put them under clean heavy books or boards to flatten them. If they are still cockled, gently mist them with water to help them relax, and put them under a weight between blotters or newspaper. If you like the cockling, you can also try removing your damp papers from the interfacing just after pressing — which usually causes them to cockle even more.

You can get some interesting results by air-drying paper. If you use a high-shrinkage pulp such as abaca or flax and let it air-dry, it will shrivel and wrinkle, creating a highly textured sheet. You can also make paper, couch, and sponge-press the sheets and work sculpturally over an armature, or cast the paper into a form (see chapter 12).

Misting gently with a spray bottle will help to relax any cockling; then flatten under weights.

Heavy books make good weights for your stacked papers while they are drying.

To loft dry sheets, be sure to rotate the sheets as you are stacking them to even out inequalities.

Exchange Drying

You can dry your sheets of paper between newspapers, cloths, blotters, or any absorbent material that will wick moisture from the paper. After they have been pressed, interleave the sheets between one of these materials, and form a stack. Put a board on top and then some heavy books or other heavy objects (buckets filled with water make good weights) to restrain the paper as it dries. Change the interleaving material daily until the sheets are dry; otherwise, they will mold instead of drying. The drying time can last from a day to a week, depending on the humidity level, the fiber, and the paper's thickness.

Loft Drying

During the advent of papermaking in Europe, a system called loft drying was developed. Sheets of paper were actually hung to dry in lofts, the top stories of buildings, where the air was considered the cleanest and warmest. The sheets were first pressed into spurs, which limited cockling as they dried.

To loft dry your sheets, you must first press them. After pressing, separate them from their felts, and pile them in stacks of four to six sheets depending on their thickness. Each pile is called a spur. Pile the sheets one on top of the other, lining them up exactly on top of the other, but rotating every other sheet so they are piled in a way that will even out any inequalities. Put a felt in between the spurs, and form a post.

Return this post to the press, and press a second time, just until you see drops of water forming at the edges of the felts. More pressure could inhibit separating the sheets when they are dry. After pressing, each spur should be stuck together. You can hang the spurs on a clothesline with clothespins. You can lay them flat on a rack or screens (see illustration on page 106). Make sure they get air from all sides so they dry evenly. In Europe, the sheets were hung from wooden poles with clothespinlike clips that did not mar the surface of the sheet. Alternatively, you can drape the spurs over plastic tubing or thick ropes (in Europe, the tradition was to hang the sheets over ropes woven from cow or horse hair coated with beeswax). Test the material you use to make sure it won't stain your paper.

If you can control the air circulation with a fan, you should try to direct the air to come in contact with the breadth of the sheets, not the edges, for the least amount of cockling. When the sheets are dry, separate the spurs by starting at one corner and peeling the sheets of each spur apart in one fluid motion. Separate the spurs in halves: if you have a spur of four, first divide it into pairs, then separate the pairs into single sheets. The sheets will probably cockle to some degree, but they can be flattened with another pressing. Make a stack of individual sheets, rotating them again, and press to flatten. If the sheets cockle a lot, you may need to gently moisten them to relax them. You can do this with a spray mister before pressing.

At Taos Paperworks in New Mexico, I saw an innovative loft-drying system. A long wooden pole approximately 6 inches in diameter was rigged to the ceiling on a pulley system. It was lowered, and spurs of papers were draped over it before it was hoisted to the ceiling, out of the way as the sheets dried. Then the pole was lowered again, and the dry sheets were removed.

Separate spurs by pulling them apart gently, using a fluid motion.

Board or Wall Drying

In China, Japan, and other Eastern countries, wet sheets of paper are brushed onto boards or walls for drying. The sheets stick to the boards and are peeled off when dry. The paper takes on the grain of the board or wall, which gives texture to the side of the sheet that was stuck to the surface. You can use almost any smooth surface for drying papers, including wood, metal, glass, Plexiglas, or Formica. You can even brush your sheets onto plaster walls, as is common in India, where the walls are heated by the sun and the sheets dry quickly.

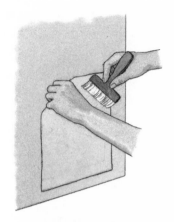

A special technique is necessary to adhere the sheet to a vertical surface, but it is easily done.

To peel the dry sheets off the board, start at one corner and peel gently with one smooth motion.

To board dry, press the sheets — they should not be quite fully pressed, or they won't stick to the boards. Lift one sheet from its felt (see Handling Damp Paper, page 104), and lay it on the drying surface. Use a rubber brayer or wide paintbrush to fix the sheet to the surface. If brushing onto a vertical surface, there is a technique you can use to adhere it while keeping it from falling to the ground. Start by brushing diagonally from the center toward the upper right corner of the sheet. While holding the upper left corner with your fingers, brush from the upper right corner to the bottom left corner. Next, brush from the center to the upper left corner, and then continue brushing from the center out to adhere the rest of the sheet. Cover the entire surface of the sheet, brushing with firm even strokes. If the sheets were pressed correctly, you should be able to brush firmly without damaging the sheet's surface. The edges must be well adhered because they will dry fast and could pop off the drying surface. If the edges curl before the sheets are dry, they will shrink more than the rest of the sheet and will be difficult to flatten.

When the sheets of paper are dry, peel them off the boards. Start at one corner, and peel in one fluid motion. If you have a problem releasing the corner, try using a razor blade to lift it, but be careful not to damage the drying surface.

AVOIDING POP-OFF

Some fibers shrink a lot when drying and will pop off the boards before they are dry. You can try applying methyl cellulose (see page 75) to the edges of the sheets as you apply them to the boards, to help them stay adhered until dry. If you are drying the sheets outside, you might try starting them in the shade and then moving them into the sun after they are somewhat dry. This allows them to dry slowly and might prevent them from popping off.

Drying Boxes

This is the highest-tech system I've seen, and all it requires is some cotton printing blotters, biwall or triwall cardboard (two or three layers of cardboard laminated together), plastic sheeting, and a box fan. The laminated cardboard is a bit costly, but this system is very efficient, yielding flat and dry papers in about 24 hours. This system can dry many sheets at a time and is designed for production papermaking.

The system works as follows: Your paper sits on blotters, which are absorbent and act as a barrier between the cardboard and the paper. The air from the fan blows through the channels in the cardboard, drying the blotters and subsequently the papers.

When purchasing cardboard to make your own drying box, make sure that the channels run in the direction in which the fan will blow air through the system. I suggest getting the cardboard as wide as the fan and no longer than one and a half times the width of the cardboard. You will need one piece of biwall or triwall cardboard and four blotters per layer, and you can stack the drying system up to the height of your fan. For instructions for building and using a drying box, see page 35.

David Reina manufactures and sells a well-designed drying box based on this type of system. It has a built-in screw press on top to ensure that the papers dry flat, which means you don't have to lift weights onto the system each time (see resource guide).

The drying box is a high-tech, more expensive way to dry your papers, but its efficiency can make it worth the expense if you plan to make paper regularly.

FINISHING

A few techniques can be performed on sheets after they are dry. Giving a finish to the paper's surface can change the absorbency and texture. The following finishing techniques are not necessary for all papers, but they may be useful for the end-use of the paper. Three common techniques are sizing, calendering, and burnishing.

Sizing

Papers that have no sizing are called waterleaf; they are generally quite absorbent, which means that any ink or watercolor applied to them will bleed. Sizing coats the fibers and makes them less absorbent. You can size your papers internally by adding size to the pulp before forming the sheets, or you can size them externally using a technique called surface sizing (see Sizing in chapter 4).

Calendering

Calendering involves repressing a stack of dry papers or passing sheets of paper between metal rollers to make them smooth. Papers are often referred to by how they were calendered. Western papers that have been pressed just once after formation are called rough. Those that have been pressed again after being separated from felts are called cold-pressed or pack-pressed. Hot-pressed papers are those that have been passed between two hot metal rollers or plates, giving them a very smooth surface on both sides. If you dry your sheets on metal, Plexiglas, or glass, you will get a smooth surface on one side, similar to the hot-pressed finish.

Burnishing

Burnishing involves rubbing the surface of the paper with some object, such as a smooth stone, to close the pores of the paper. An early burnishing tool was an agate stone, used in India to give papers a smooth writing surface. Other items such as leaves, plastic tubing, and glass bottles have also been used to burnish papers. In 16th-century Europe, a glazing hammer was developed to burnish sheets of paper mechanically.

Many objects can be used as burnishing tools, as long as they are smooth and heavy enough to close the pores of the paper.

Papermaking
Techniques
and Projects

CHAPTER 6

PULP PAINTING

Paper pulp can be pigmented and used as a medium to paint on the surface of a wet sheet of paper — the wet base sheet is similar to a painting canvas. Layers of pulp can be applied and built up on the freshly made sheet of paper. When pressed, the layers flatten and bond, becoming one uniform surface.

Pigmented paper pulp has a different consistency than paint. It can be watered down so that you can pour it or spoon it. You can manipulate it by squirting it through a bottle, turkey baster, or syringe. If it is refined enough, you can even work it with a brush. As layers of pulp are applied to the base sheet, they can be misted with water to create varying intensities of color, scraped through to bring up colors that lie underneath, and manipulated in a variety of ways.

PREPARING YOUR PAINTING PULP

Linen rag, cotton rag, cotton half-stuff, or abaca beaten for 4 to 8 hours in a Hollander beater makes an ideal painting pulp. Although you can paint with any pulp, one that is overbeaten in a Hollander has several advantages:

- Overbeaten pulps have more bonding surfaces opened up, so they can accept more color to acquire intense hues.
- The longer the pulp beats, the more hydration occurs, which makes the pulp "greasier" (it flows more consistently).

BASE SHEET PULP

If you are painting with overbeaten pulps, I recommend you use a base sheet that is not made from overbeaten pulp. Overbeaten pulp is more difficult to form sheets with, since it's so slippery and drains slowly. In addition, it shrinks more as it dries and will be more difficult to dry flat.

◆ Overbeaten pulps are fine enough to be worked with a brush, squirted through a syringe, or moved around by being sprayed with a mister once they are on the surface of the base sheet.

If you do not have access to a Hollander beater, you can order prepared painting pulps from papermaking suppliers. Rick Hungerford, a papermaker who specializes in pulp painting and teaches around the country, sells his overbeaten cotton half-stuff, which he refines for pulp painting (see resource guide).

Again, you can paint with any pulp; it's just that underbeaten pulps don't accept the intense colors that overbeaten pulps do. They also tend to be less fluid, though this can be remedied by adding formation aid or methyl cellulose to lubricate the fiber. With any pulp, you will need to experiment with the water-to-fiber ratio.

Many artists prepare a palette of pulps before painting. After preparing the pulp, divide it into containers and pigment each batch (see Pigments and Dyes in chapter 4). Add retention aid, and then use one of the painting techniques described below. If you plan to mix a batch of painting pulps to last awhile, you can put them in resealable plastic bags or plastic yogurt containers, and store them in the refrigerator between uses.

PULP PAINTING TECHNIQUES

Following are some methods I've seen for painting with pulp; they are by no means the only ones. After you've tried a few techniques, you can work solely with one, or combine several to create your own pulp paintings. All these techniques involve making a "base sheet" of paper and couching it, and then painting on top of the sheet.

For each technique, sheets should be pressed and dried after painting with pulp. I do not recommend board-drying pulp paintings if you use overbeaten pulps — they tend to stick to surfaces and could be difficult to remove.

TRUE COLOR

Getting a true black paper is difficult, particularly with an opaque fiber such as cotton — since you are counteracting the whiteness of the fiber, you will almost always get a dark gray instead of a true black. When coloring natural plant fibers, remember that the original color of the fiber will affect the results. Keep notes so that you have a record of what works.

Experiment with different kinds of squirt bottles, syringes, and basters.

Cookie cutters are a simple but very effective way of applying pulp to your paper to make patterns.

Use string to draw shapes on the paper and then use the string to contain the pulp.

Squirt Bottles, Basters, and Syringes

You can use a variety of implements to apply pulp to the base sheet. Plastic squirt bottles, such as the ones used for mustard and ketchup in restaurants, work well and come in a variety of sizes from papermaking suppliers. Dental syringes work very well for fine and controlled lines. Turkey basters are another possibility, although I find I have less control with them. You can clip the nozzles on any of these implements to create thinner to wider painting lines for application.

Make sure the pulp is diluted enough to let it flow through the spout of your squirting tool. If you use a bottle, shake it often to mix up the fiber and water. You may need to add water from time to time; the water tends to flow out faster than the pulp, sometimes leaving only pulp in the bottle.

To aid in fluidity, you can also add some formation aid (see Formation Aid in chapter 4) or methyl cellulose to the pulp and water, then mix the solution well by shaking.

Cookie Cutters

A simple way to apply pulp to certain areas of paper is to use cookie cutters as stencils. Set a cookie cutter on top of the freshly made base sheet of paper, and press it gently onto the surface. Spoon, squirt, or pour a watered-down pulp into the cookie-cutter shape (and/or around the outside of the shape). The pulp will be contained by the cookie cutter. Let the water in the pulp soak into the sheet of paper before lifting the cookie cutter — otherwise the pulp will spill out over the edges, and you'll lose the shape. You can overlap shapes and colors working in this manner, or just work in certain areas.

Try creating your own shapes using Styrofoam, interfacing, or Mylar instead of cookie cutters. You can also draw shapes with string and use the string as the line to contain the pulp.

Other Ready-Made Stencils

With some creative thinking, you can find interesting stencils in your own home. Lay plastic fluorescent lighting grids, milk crates, or metal heat and air conditioning grates on top of the base sheet; then apply colored pulps between the lines on the grids. Leave the grid on the sheet until the colored pulps have sufficiently drained.

MARBLEIZING PAPER

Paul Wong, artistic director at Dieu Donné Papermill in New York City, has developed this recipe for making marbleized paper:

You Will Need:
> Base sheet
> Various pigmented pulps in containers (8-hour beaten linen or cotton rag, half-stuff, or abaca; or cotton linters, diluted with a thick solution of methyl cellulose or some formation aid to make the pulp more viscous)
> Plastic cups
> Brush or other implement to swirl colors
> Mould and deckle (or deckle box)
> Base sheet
> Stencils (optional)

1. Dilute pigmented pulps to a creamlike consistency with water.
2. Dip a plastic cup into two or three different colored pulps, collecting a bit from each. With a brush, stir the contents of the cup once so that the colors swirl but don't blend.
3. Pour the swirled pulp onto the surface of the papermaking mould in a random fashion. Work on a wet floor or over a tub that will collect the water as it drains through your mould.
4. Repeat step 2 until the desired area of the mould is covered.

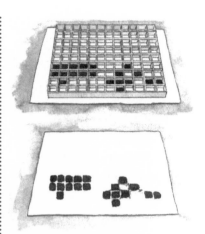

Many different objects can be used as stencils for painting with grids.

step 3

step 5

5. Couch the marbleized sheet onto your already formed base sheet. This will give the poured sheet a substantial base, especially good when you are using high-shrinkage pulps.

ANOTHER MARBLEIZING TECHNIQUE

These sheets actually look a bit like the traditional marbled papers that can be found in old books. This technique was contributed by papermaker Robbin Ami Silverberg of Brooklyn, New York.

You Will Need:
> Pulp for base sheet
> Mould and deckle (or deckle box)
> No-see-um netting, cut to the size of the mould
> A few pigmented pulps in containers (overbeaten abaca, flax, or cotton)
> Plastic cups
> Formation aid
> Chopstick, paintbrush tip, pencil, or other fine-tipped tool

TRY THIS VARIATION

Place shaped stencils on the papermaking mould before pouring painting pulp, to confine colors to specific areas or to create a marbleized image.

1. Make a base sheet, and couch it onto a felt.
2. Place the no-see-um netting on top of the mould, and then set the deckle on top (the netting slows the drainage time).
3. Dip a cup into one of the pigmented pulps, mix in some formation aid (1 or 2 tablespoons), and pour a line of pulp onto the mould.
4. Repeat with other colors to cover the mould with lines of pulp.
5. Drag a chopstick or other fine-tipped tool through the pulps in a direction perpendicular to the lines you have already created.
6. Continue dragging through the pulped lines at regular intervals of approximately 2".
7. Allow the sheet to drain thoroughly, and then couch it onto your base sheet.

step 5

Fine-Line Stenciling

This is another of Paul Wong's techniques. It can be used to create a line drawing by slitting frosted Mylar and then "print" an image with colored pulp. Start with a simple line drawing. You can then develop several layers of cut Mylar and "print" an edition in several colors.

Frosted Mylar can be used for fine-line stenciling.

You Will Need:

Line drawing	Pulp for base sheet
Frosted Mylar	Pigmented pulp
Tape	Spoon
Cutting knife	

1. Create the Mylar stencil: Place your line drawing underneath the Mylar, and tape both in place, or trace your drawing onto the Mylar. Using a cutting knife, slit the Mylar, leaving small ⅛" pieces intact every few inches so that the stencil does not fall apart.

step 1

2. Make a base sheet, and couch it onto a felt.

3. Set the Mylar stencil in position directly on top of the base sheet.

4. Spoon the pigmented pulp onto the slits in the Mylar stencil. Use the spoon to gently massage the pulp in a circular pattern all over the cut lines of the stencil. Take care not to gouge or catch the edges of the stencils with the spoon. Add formation aid or methyl cellulose if the pulp is not easily manipulated.

step 4

5. Carefully remove the Mylar stencil, which can be reused.

6. Press and dry the sheets (see chapter 5).

TIP

Use a short pulp, such as cotton, if you want the line drawing to feather, or a tight, overbeaten pulp, which will not feather the line as much when the pigment seeps through the stencil — it will deposit a fine hairline impression of the image.

Blockouts

Interfacing or Mylar can be used to block out areas on the base sheet when you are applying color.

You can cut shapes out of nonwoven interfacing or Mylar to block out areas on your base sheet when applying color. (I prefer interfacing because it soaks up any pulp that hits it, whereas the pulp and water mixture floats on top of the Mylar).

Simply cut out shapes and lay them on top of the base sheet. Then apply a wash of pulp — mixed with formation aid — to the sheet with a small plastic cup or spoon. When done, peel up the interfacing or Mylar to reveal the unpainted shapes. You can repeat this to build up multiple layers. Rinse your Mylar or interfacing between layers to prevent getting unwanted pulp in areas of your work.

OTHER TECHNIQUES

Using a Mister

After applying colored pulps to the base sheet, you can manipulate them to get interesting three-dimensional effects by spraying the pulp with an atomizer or spray nozzle attachment from a hose. This works best with overbeaten pulps. Use pieces of nonwoven interfacing if you want to block certain areas from misting.

Layers of Pulp

If you lay down several layers of colored pulps, you can actually break through the layers and reveal other colors underneath. You can do this by using a mister, forcing water through a syringe, or gently scraping through the layers with a blunt implement, such as the back of a paintbrush or syringe.

Painting with a Brush

To pulp-paint with a paintbrush, use overbeaten pulp mixed with lots of formation aid, which will allow a fluid brush stroke.

CHAPTER 7

LAMINATING AND EMBEDDING

Working on top of a freshly couched sheet of paper, you can laminate several sheets one on top of the other, or collage other materials onto the sheet. You might try making a two-toned sheet of paper by laminating one color or fiber to another; when pressed and dried, the fibers will bond and become one sheet of paper. Or, collage items such as photographs, strings, and fabric directly onto the surface of a sheet when it's still wet. When laminating multiple sheets, you can also embed items between the sheets. (See "Lining up Sheets" on page 124.)

SIMPLE LAMINATING

Any kind of pulp can be used for laminating; you can experiment with translucent and opaque fibers. After making a sheet of paper, transfer it to a couching surface. Using a different pulp, make a second sheet of paper, and couch it directly on top of the first. You can vary the outcome by couching the second sheet at an angle to the first, or just overlapping part of the first sheet; when pressed and dried, the fibers will bond where connected.

This is also a way to make a large sheet of paper if you have a small mould. Overlap the sheets by at least 1 inch to provide a strong joint. Obviously, the sheets will be thicker where they overlap.

If you have only a small mould, overlap two sheets to make one large sheet of paper.

LINING UP SHEETS

In order to couch one sheet of paper directly on top of another, you will need a guide or jig for lining up your sheets. Follow these steps:

1. While the couching surface is still dry, lay the mould face down in the middle of it. Using ¼" wide strips of waterproof tape that are 2" to 3" long, make a couching guide by marking two adjacent corners — preferably on a board or tray outside your papermaking area, so that you can use the guides to line up more than one sheet.

2. Make a sheet of paper and couch it, laying one side of the mould down along the guides.

3. Make a second sheet of paper and couch it on top of the first sheet, using the guides to line up the two sheets.

COLLAGING

Use methyl cellulose glue for collage items by applying with a brush or dipping the object into a container of the glue.

You can collage directly on top of a freshly made sheet of paper with items such as photos, drawings, thread, string, plant materials, or even feathers. Most materials will not bond to the sheet of paper without the aid of an adhesive. Methyl cellulose glue (see chapter 4, page 75) is an archival, clear-drying adhesive that works well for this application. You can brush it onto the back of the items you are collaging, or submerge the items in a container of methyl cellulose (the latter method works well for items that are difficult to brush with glue — a feather, for example).

EMBEDDING

You can sandwich items between two sheets of paper to create interesting textures or add structure to a sheet. Use a translucent fiber on one side (or both) to embed an object such as a photograph or a leaf, to make it visible or slightly hidden. Embed a piece of lace, plastic mesh, or a handful of seeds to make a textured paper. Or, sandwich pieces of string or wire to make structural sheets that can be bent or manipulated.

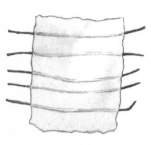

Process Steps

To embed an item, make a sheet of paper in any pulp (this will be the backing paper for your images). Place a leaf, photo, or other memento on the sheet. Make another sheet using a translucent pulp (such as kozo or abaca) and couch it, lining it up with the first sheet and sandwiching the memento. Press and dry the paper.

In general, you do not need to use an adhesive when embedding, because you are actually trapping the objects between two sheets, which will bond. However, if what you want to embed covers a large surface area — perhaps a photograph or a large piece of fabric — I recommend applying methyl cellulose to both sides of the item. If you don't, the item may become loose between the sheets of paper.

To sandwich wire, make a sheet of paper and couch it. Set up a jig for lining up the next sheet. Lay strips of wire on the freshly couched sheet. Now make a second sheet of paper, line it up, and couch it on top of the first sheet. When the paper is dry, the wire will give it support and allow manipulation by bending. Try using a high-shrinkage pulp — the pulp will shrink and bend the wire as it dries (if air-dried) to create interesting forms.

Note: If you are embedding wire, use galvanized steel or copper to avoid rust. Refer to Casting (chapter 12) if you want to create paper with relief (with items thicker than ⅛ inch).

Many different kinds of objects can be embedded between two sheets of paper.

Sandwiching wire between your paper sheets allows you to bend it.

MAKING A LAMINATED WIRE LANTERN

Make these simple lanterns to decorate a tabletop or windowsill. Since wire is the internal structure for these pieces, you can bend and manipulate the paper when it is dry to create any of several interesting shapes. Embedding washers or other types of fasteners in the paper also gives you options for joining small sheets together to create unique forms. When making the paper, create an intriguing pulp with inclusions (see chapter 8) or watermarked designs (see chapter 11); or embellish your sheet with a pulp painting (see chapter 6) or collage to make your lantern unique.

You Will Need:
> Prepared pulp
> Mould and deckle
> 18-gauge galvanized steel or copper wire (any rustproof wire
>> will do), available at craft stores
> Glass votive holder and candle
> Small stainless steel washers (or steel washers coated with
>> polyurethane)
> Nuts and bolts to fit washers

Note: For this project, I made small sheets of paper (5" x 6"), which I joined together with nuts and bolts. I like the contrast of combining the warm, soft beauty of the paper with cold, hard, metal pieces. You could join the pieces by stitching, stapling, or gluing them together if you don't want to embed washers. You can also extend the lengths of wire beyond the edges of the sheets so that they can be twisted together to assemble the lantern.

Your sheets of paper should be at least 5" tall, and the diameter of the final lantern should be at least 5", so that you can fit a protective glass votive holder and candle inside.

1. Make a sheet of paper, and couch it. Set up a couching guide for aligning the next sheet (see page 124).

2. Cut several lengths of wire to the length of the sheet of paper (or longer, if desired). Straighten them as much as possible, and place the lengths at approximately 1" intervals. If you are using washers, place one in each corner, approximately ¼" in from the corners in both directions.

3. Make another sheet, and couch it directly on top of the first sheet, embedding the wires and washers between the sheets.

4. Repeat steps 1 through 3 to make at least four sheets for the lantern.

5. Press and dry the sheets.

6. When dry, assemble the lantern using nuts and bolts, by stitching it, or by twisting the wires together to form a seam on the inside or outside of the lantern.

7. Place a votive candle in its protective holder, light it, and carefully set the lantern over the candle. Now step back, and enjoy your creation!

Variations

◆ Use hardware cloth or chicken wire instead of lengths of wire.

◆ Create various shaped lanterns or screens by making more paper panels.

step 2

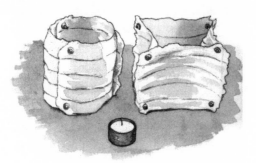

MAKING A LAMINATED BOOK COVER WITH POCKETS

Laminate sheets of paper together to create a book cover. By using a piece of interfacing to block where the lamination occurs, you can make pockets in the cover, into which you can slip a signature (a group of pages, in this case formed by folding several sheets of paper in half).

Book covers get worn as they are opened and closed, and they also receive a lot of handling, so I recommend using a pulp that will hold up to wear and tear (at least 25 percent abaca, cotton linter, or rag pulp).

Your book cover can be whatever size you like. I simply used an 8½" x 11" mould and deckle to make this cover, which folds in half to 5½" x 8½". The signature is made of 8½" x 11" sheets of paper, which are also folded in half. If your mould and deckle are a different size, you will need to do some planning before you begin. Or, if you wish to make your cover a certain size, you can make a substitute deckle to make that size. Books are usually planned from the inside out, so you might want to assemble the signature first to determine what size the cover should be. I like to make my cover at least ⅛" bigger all around, so that the signature fits nicely inside.

Making the Cover

For this technique, you will need to set up a guide for lining up the sheets (see page 124). You can apply any papermaking technique (such as pulp painting, inclusions, embedding) to embellish the cover.

You Will Need:
Prepared pulp
Mould and deckle
Interfacing cut to 7½" x 10" (one for each cover you
    plan to make)
Ruler
Dental syringe
Tape for aligning sheets

TIP

You can use the cover indefinitely — when you fill up your pages, simply slip them out and slip a new signature in.

1. Make a sheet of paper that measures 8½" x 11", and couch it.

2. Lay a piece of interfacing or plastic sheeting on top of the sheet, centering it in all directions.

step 2

3. Make a second sheet of 8½" x 11" paper, and leave it on the mould. Now you are going to remove a strip of pulp from the middle of the sheet, to create the two inside covers for your book. You can do this by eye, or use a ruler to make two marks (with a pencil or tape) on the top and bottom edges of the deckle. Then, lay your ruler on the deckle between two of the marks, and use it as a guide for drawing the cutlines with the syringe. Move the ruler to the next set of marks and repeat. You will end up removing about a ¾" strip of pulp. Using a dental syringe filled with water, and starting 5⅛" in from the left edge of the sheet, squirt a line from the top to the bottom of the sheet. Squirt another vertical line 5⅞" from the left edge.

step 3

4. Using the back of the syringe, gently draw the strip of pulp toward the center, pick it up, and remove it.

step 4

5. Line the sheet up with the previously couched one, and couch it on top of the sheet and the interfacing or plastic.

6. Press and dry.

7. When the sheet has dried thoroughly, gently remove the interfacing to reveal the pockets. (See illustration on page 131.)

Variations

◆ Embed closure materials (ribbon, string) in the front and back edges of the sheets.

◆ Cut the spine area with freeform lines to create an interesting edge.

MAKING A SIMPLE SEWN SIGNATURE

There are many ways to make signatures for books; this one is called the five-hole pamphlet stitch.

You Will Need:

> 4 sheets of 8½" x 11" paper
> Card stock cut to 10" x 6"
> Bookbinding thread
> Pushpin, awl, or potter's needle
> Needle
> Bone folder

1. Fold the four sheets of paper in half (lengthwise). Nest one inside the other to form a signature.

2. Use a bone folder to score the 10" x 6" piece of card stock and fold it in half (lengthwise). Trim the corners at slight angles.

step 3

3. Open the signature, measure, and mark the following five points in the fold: 1" down from the head; 3¼" from the head; the center; 1" up from the tail; and 3¼" up from the tail.

4. Open the card stock, and center the signature inside it.

5. Open the signature flat on a tabletop with the inside facing up. Use a pushpin, awl, or potter's needle to punch holes through the signature and card stock at the five markings.

6. *(a)* Thread the needle, and start sewing through the center hole on the outside of the piece. Leave a 1" tail of thread outside. Continue sewing by coming out of the hole directly above the center hole, and then sew back in through the top hole. *(b)* Bring the needle out of the second hole, then carry down and stitch through the fourth hole to the inside. *(c)* Bring the needle back to the outside through the bottom hole. Finish the bottom half by sewing into the fourth hole and *(d)* back out through the center hole. When sewing

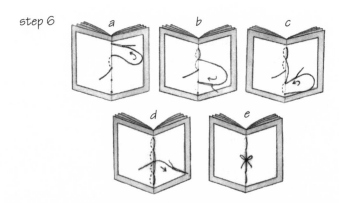

step 6   a   b   c   d   e

this last stitch, try not to sew through the existing thread, and keep it on the opposite side of the long stitch from the tail. *(e)* Tie a bow or knot, and trim off any excess thread, leaving tails about ½" long.

7. Slip each side of the card stock into a pocket on the cover to assemble your book.

step 7

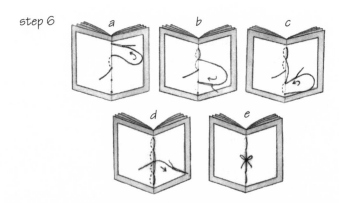

CHAPTER 8

DECORATING WITH INCLUSIONS

Adding items (often referred to as inclusions) to your papermaking vat is one of the simplest ways to make unique and intriguing papers. You can toss almost anything that floats into your vat of pulp just before making paper. Decorative elements should be small and light-weight, so they don't sink to the bottom of the vat; they should also be flat, so they don't cause problems when pressing. Plant fibers, flower petals, or small swatches of fabric are just a few possibilities — the items just need to float amid the pulp so that they get picked up on the mould when you form a sheet.

Some interesting inclusions I have seen in papers include coffee grounds, tea leaves, shredded money, and bits of bird nests. If you have small amounts of plant fiber or leftover bits of pulp, simply toss them into a vat of another base pulp to create new and decorative papers.

BASIC INCLUSION PROCESS

To make paper with inclusions, simply prepare the pulp as you normally would, and add the decorative material just before making sheets. You can add most materials to the pulp during the last few minutes of beating, to mix them thoroughly into the pulp. Otherwise, you can toss them right into the vat — just make sure you stir the pulp so the material is evenly distributed.

You can add almost anything to your pulp that is small and will float.

TIP

Many decorative items are not archival, so take this into consideration if you want your paper to last for hundreds of years.

Adding inclusions to the pulp will result in their random overall placement in the sheet. If you want your objects placed in particular areas on the sheets, you should lay them onto the sheet after it is made (see chapter 7 for details on collaging and laminating).

FLOWER PETAL PAPER

Papers with flower petal inclusions are popular, pretty, and easy to make. Pick the petals from their stems, and toss them into your vat. Many flower petals bleed when they get wet, and this discoloration can spread into the sheet of paper, sometimes creating unattractive stains. You can blanch most petals to prevent bleeding, but in some cases, all the color will be removed, and you will be left with colorless petals — which are nonetheless quite beautiful. Keep records, so that you know what works the next time around.

You Will Need:

Flower petals	Prepared pulp
Stove	Strainer
Pot	Papermaking equipment

Blanching Flower Petals

Blanching the petals in hot water before adding them to the vat can prevent bleeding, and it will also hydrate the material so it mixes into the pulp more readily. Follow these steps:

1. Gently remove the petals from their stems.

2. Bring a pot of water to boil, add the petals, and blanch for 5 minutes.

3. Strain out the petals, and add them to the vat of pulp. You can vary the amount of petals to make a subtly or highly decorative sheet.

4. Make a sheet of paper, and couch it.

5. Press and dry your sheet.

TIP

When making sheets, remember to stir the pulp continually in the vat to ensure even distribution of the decorative items.

RAGS AND RICHES

Ron and Jennifer Rich discovered handmade paper at a gallery while looking for a wedding-photo album in 1989. Along with the album, they found a how-to book on papermaking: Faith Shannon's *Paper Pleasures*. During their month-long honeymoon on the Oregon coast, using the book as a guide, they threw sea grasses and pods into a secondhand blender and began making paper. Ron, who also made stringed instruments, constructed simple moulds and deckles.

At first, the Riches made journals and cards to give as Christmas gifts to family and friends. Soon they started producing and selling their creations at the Eugene, Oregon, Saturday Market, a weekly craft fair. They also studied hand papermaking and learned that decorative floral handmade papers are made in France. That piqued their interest, and they took a bicycle trip to visit the old mills there, some of which were still producing papers. At Moulin de Larroque in Couze-et-St.-Front, the only remaining hand-papermaking mill of 13 that had once produced papers there, a first-hand view of the French papermaking process inspired them to refine their own floral papers and upgrade their equipment. The rhythms of the vats, the smell of damp paper, and the store rooms stacked with freshly made sheets of paper inspired them to run a full-time hand paper mill.

The Riches began producing papers with a base pulp of cotton linters (later cotton half-stuff) and inclusions, such as calendula, carnation, and bachelor-button petals; fern leaves; and grasses. They began selling at craft fairs throughout the western United States. Production increased, and Ron continued to build more sophisticated equipment. He built a Hollander beater from scrap wood and plastic, as well as several moulds and deckles, a press, and a couple of drying boxes. Several years later, the Riches invested in a professional Reina Hollander beater.

When traveling to craft fairs, Ron and Jennifer stopped at retail stores and peddled their products, adding wholesale accounts to their business. In 1994, they attended their first trade show in New York City. Orders generated there enabled them to begin hiring staff.

The Riches lived on Vashon Island, Washington, for a few years and developed a relationship with Tabula Rasa, a letterpress printer in Seattle. After several orders for letterpress-printed handmade paper invitations, the Riches purchased their own press and produced the whole package themselves. They found a hundred-year-old old-style Chandler and Price letterpress, and Jennifer learned to print. At first they set lead type by hand; eventually they switched to photo-polymer plates, which can be generated from camera-ready art.

The Riches now focus their business on wedding invitations but continue to produce photo albums, journals, guest books, notecards, and writing papers. In 1998, they moved their company, Oblation Papers and Press, to Portland, Oregon. There they operate their production studios, which can be viewed from their retail store.

Selecting Flowers

Of course you can let your flower petals bleed — you might find you like the effect. But keep in mind that several types of flower petals do not bleed and therefore do not require blanching. The list at the left I found on the Internet (see resource section); I presume that these petals have been tested but cannot guarantee them myself.

NONBLEEDING FLOWER PETALS

Bachelor button
Bee balm
Cornflower
Carnation
Chrysanthemum
Fern
Geranium
Golden glow
Larkspur
Lobelia
Marigold
Monkshood
Obedient plant
Pansy
Phlox
Statice

OTHER CREATIVE INCLUSIONS

Other decorative items can give your paper a unique quality. Options include confetti, strands of thread, leftover plant pulp, glitter, and leaves — use your imagination! You can also use up a bunch of old pulps by adding a coagulant (available from papermaking suppliers) to each pulp and then mixing them together in a vat. The coagulant will make small clumps of each color form, and these will mix together to make unique sheets. The clumps of fiber will maintain their indidvidual colors rather than all mixing together to form a uniform brown or gray sheet.

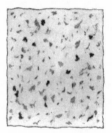

Using a coagulant causes colors to clump together, making unique patterns.

CHAPTER 9

EMBOSSING

You can create textured sheets of paper by pressing objects against the surface right after sheets are formed. This is known as embossing. The texture of a sheet will vary, depending on the type of felt used for couching. If dried against a smooth surface, the resulting sheet will be smooth. Sheets dried on wooden boards will pick up the grain of the wood. To achieve the finest embossed detail, press an object against a freshly made sheet of paper, and then dry the two together under a weight, removing the embossing item only after the sheet is dry.

GETTING STARTED

Low-shrinkage pulps provide the greatest definition during embossing. Even a knotty, lumpy cotton linter or recycled pulp will work — you just need a consistent pulp. If you use a drying box (see drying techniques on page 113), you can try other high-shrinkage pulps, because the mechanics of the drying box produce dry papers in about 24 hours, with almost no shrinkage. If you don't have access to a drying box, though, be sure to keep in mind that, in general, shrinkage results in loss of detail.

BASIC EMBOSSING

The items you choose to emboss should be relatively flat — no thicker than ⅛" — and nonbreakable. (Try casting, explained in chapter 12, if you want to emboss thicker items.) I recommend using thick couching material such as felts when embossing, to absorb the impression of the added items when pressing.

Some suggested embossing items are listed below. Keep in mind that many metals rust when they come into contact with wet paper. This can create interesting effects, but if you want to avoid rust, use copper, brass, or galvanized metals.

You Will Need:
Textured fabric
Plastic
Rubber
Leaves
String
Coins
Wire
Metal grating

step 2

1. Make a sheet of paper, and couch it onto a couching surface.
2. Place the items you wish to emboss directly on top of the freshly couched sheet.
3. Place another felt on top of the sheet and the embossing items. If you plan to make a post of paper and can see the embossing items through the felt, I recommend adding another felt layer as an additional cushion. (This is not necessary if you are embossing something thin, such as fabric.)

step 3

4. When you finish forming the post, stack an additional felt on top, and press the sheets.

5. Transfer the sheets to a drying system. For best results, leave the embossing items on the surface of the sheets, and let them dry in place. Restraint drying (see page 110) between blotters in a drying box will result in a well-embossed sheet.

step 5

 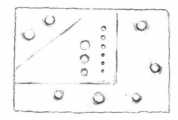

Pressing Thicker Items

In some cases, you can actually use a press to press items that are ¼ inch to ½ inch thick. To do this, keep in mind that the sheet of paper you are forming must absorb the embossing object so that it does not tear during pressing. This might require pouring a thick sheet of paper onto a mould to form a sheet that is almost an inch thick. Make sure you let it drain sufficiently before couching. When pressing, stack several felts and then a piece of foam to cushion the thickness of the embossing item. Press carefully to ensure that you do not break the item.

For some embossed items, you can use a vacuum table to suck out excess water and draw your sheet to the freshly made sheet of paper. This method will also save time spent sponging.

MAKING EMBOSSED OR DEBOSSED MATS FOR PICTURE FRAMES

Artist Mary Leto developed this project for children, but it is great for "kids" of any age (adults included). Create unique embossed mats using common household materials as molds. You can also use this technique to create book covers and greeting cards.

You Will Need:

Scissors
Styrofoam (use clean large meat trays or packing
    Styrofoam; paper tends to stick to porous types)
Ballpoint pen or other blunt-tipped tool (knitting needle, pencil)
White glue
Prepared, low-shrinkage pulp (such as recycled or cotton linter)
Stiff paintbrush
Sponge
Cutting knife

step 3

a

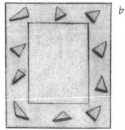

b

1. Cut the Styrofoam to the desired outer dimensions of the frame (6" x 8" works well to frame a standard 4" x 6" photograph).

2. Use a blunt-tipped tool to mark a 1¼" border from each edge of the Styrofoam. This will mark a 3½" x 5½" window, which will be cut out of the mat to display your photograph. (Variations: you can make the window any size, make more than one window, or make the window off-center for creative frames.)

3. Use a blunt-tipped tool to impress a line drawing or design into the Styrofoam border *(a)*. Press firmly so that you can see an indentation. Or, cut shapes from another piece of thin Styrofoam (approximately ⅛"), and glue them to the border *(b)*. This will make a debossed design (indented rather than raised) in your mat. When gluing, make sure no glue seeps out from underneath the pieces and is exposed, because it will stick to the wet paper pulp.

4. Make a sheet of paper (if you are using the cut-shape technique, make sure it is thick enough to compensate for the thickness of the shapes), slightly larger than the piece of Styrofoam. Couch and press it lightly.

5. Transfer the sheet, and lay it on top of the Styrofoam design. Use a stiff paintbrush or a clean, damp sponge to stipple (gently prod) the paper and press it into the design. You can work directly on the sheet of paper, or place a thin piece of interfacing on top of the wet sheet. After the paper has formed to the shape of the Styrofoam, place a piece of interfacing on top of the sheet, and sponge-press to remove as much water as possible.

6. Air-dry the mat. If the paper starts to shrink away from the Styrofoam as it dries, set a plastic bag filled with something heavy (sand or beans) to weight down the edges of the mat with the Styrofoam still beneath.

7. When the paper is dry, remove it from the Styrofoam, trim the outer edges (if desired), and cut out window/s for your photos, using a mat knife.

TIP

You can accentuate your design by rubbing over the surface with a bit of acrylic or metallic paint (available in art supply stores), which will highlight the embossed areas.

---

step 5

step 7

CHAPTER 10

MAKING SHAPED SHEETS

You can vary the shape of your paper by designing shaped deckles that control where pulp flows onto the mould. You can make circles and stars, or even divide the mould surface so you make more than one sheet at a time, for small sheets such as cards or stationery.

MAKING SHAPED DECKLES

There are several materials on the market that you can use to create shaped deckles. All of these will enable you to make uniquely shaped papers. Some papermaking suppliers (see resource list) also sell ready-made shaped deckles for making envelopes and other shaped sheets.

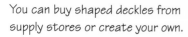

You can buy shaped deckles from supply stores or create your own.

♦ **Artcor** is my favorite, though it can be somewhat hard to find. Lightweight, like foam core, it's made entirely of plastic and is therefore waterproof.

♦ **¼ inch Styrofoam** is very cheap and is available in large sheets at lumberyards. Perfectly suitable for shaped deckles, it has all the advantages of artcore but is more fragile.

♦ **Foam core** will work, but it won't last long because it has a paper surface. You can cover it with plastic tape to waterproof the surface and increase its durability.

142

◆ **Masonite,** or other thin, lightweight wood or Plexiglas, can be cut to a particular shape with a jigsaw. I would do this only if you are going to use this deckle a lot. You should polyurethane any wood so that elements in the wood don't stain your paper.

You can cut the shaped deckle to fit either the inside or the outside dimensions of your regular deckle. If you cut it to the inside dimension, use strapping tape or other waterproof tape to attach it into your regular deckle. Tape it to both the top and bottom sides of the deckle at all points of contact.

If you cut the shaped deckle to the outside dimension of your regular deckle, sandwich it between the mould and the regular deckle when forming sheets *(a),* or use the shaped deckle in place of the regular deckle *(b).*

Use tape to attach your shaped deckle to the inside of your regular deckle.

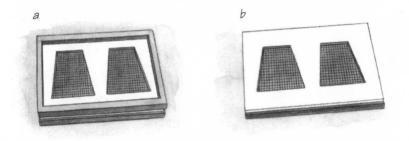

a

b

Once you've created a new deckle, simply form sheets and couch them as you normally would. It is a good idea to create a guide for lining up your sheets as you build your post, so that they're lined up evenly for pressing (see page 124). Also, pay attention to which way you couch, making sure that you don't flip the mould and misalign your sheets. If your shaped deckle is divided into several sheets, you will obviously spend more time handling the sheets during the drying process.

TIP

If you are having problems with pulp slipping under the edges of your shaped deckle, try taping tiny weights to the top of the deckle (old lead type or small magnets work well) to keep it from floating up as you form sheets of paper.

HOW TO MAKE A PLASTIC DROP CLOTH DECKLE

If you are making just a few shaped sheets of paper, you can cut the shape out of a lightweight sheet of plastic, such as a drop cloth. Cut a sheet of plastic to the outer dimensions of the mould; then cut a simple shape in the center of the plastic. Wet the plastic, place it on the mould's surface, and smooth it out. Put the deckle on top, and make a sheet of paper. The sheet of paper will probably drain slowly because the plastic covers areas of the mould surface. Try carefully lifting the deckle to allow some of the excess water to slip out the sides. After the sheet has sufficiently drained, remove your real deckle, and then carefully remove the plastic deckle by gently slipping it out from underneath the shaped pulp. Couch the sheet.

PAPERMAKER'S PROJECT

CUTTING SHAPES IN WET PULP

Here is another way to create unique shaped sheets. I learned this from Richard Hungerford, an artist in Iowa who uses water forced through a dental syringe (available from A. Leventhal & Sons, Inc. — see resource guide) to cut lines in freshly formed sheets of paper.

You can use this technique to create holes in a sheet of paper or to create a thin line that can be torn later, functioning similarly to a perforation. Squirting a straight line can be a little tricky; try resting a straightedge on top of the deckle as a guide.

Rest a straight edge on the deckle to help you squirt a straight line with a dental syringe.

You Will Need:

Pulp
Mould and deckle
Dental syringe or other squirting device

1. Make a sheet of paper, and leave it on your mould, prior to couching.

2. Fill a dental syringe with water, and squirt the outline of a shape onto the freshly formed sheet. You need to squirt with force to break the fibers apart.

3. After squirting the line, gently peel the unwanted pulp away from the line. If you have a lot of pulp, you can use the part you're removing somewhat like a kneaded eraser to attract bits of pulp and clean up the lines. Take care not to squeeze this pulp too hard — leave it loose and you'll be able to dump it right back into your vat. Be sure to mix the vat well before forming another sheet.

OTHER WAYS TO MAKE SHAPES

There are several other simple ways to make shaped papers. You can use cookie cutters to make controlled shapes, or use your hands or pour pulp to create less-controlled abstract shapes.

NO LIMITS

Arnold Grummer, a papermaker who sells a series of instructional papermaking kits, developed a unique way of making paper using old tin cans. He has a variety of products on the market and has written a book called *Tin Can Papermaking* (see reading list).

You can use cookie cutters, or cut a stenciled shape from Mylar, interfacing, or Styrofoam, and place it on the mould. Pour pulp into the shape, and then couch the shape — by itself or onto a base sheet.

step 2

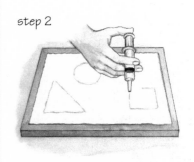

step 3

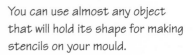

You can use almost any object that will hold its shape for making stencils on your mould.

HAND-FORMING

A simple but crude way to make a shaped sheet is to form the sheet on the mould and then remove pulp by pulling it away with your hand. Remove the pulp before couching the sheet, when it is easiest to peel away. You can lay a paper or plastic stencil on top of the freshly formed sheet to guide you in where to remove pulp.

You can create shapes on your mould by simply pulling the pulp away with your hand.

POURING SHEETS

Another simple way to make a shaped sheet is to just pour it freeform on top of the mould. After it drains, couch the sheet onto a felt or a base sheet.

Papermaker's Project

MAKING MULTIPLE CARDS AND ENVELOPES

You can make two or more cards in one dip using a shaped deckle with windows cut into the shape of cards. You can also create unique envelopes in handmade paper. This project is based on using a mould and deckle that creates papers measuring 8½" x 11", but it can be adapted for any size mould.

You Will Need:
    Mould and deckle
    Deckle material (see page 20)
    Ruler
    Cutting knife
    Waterproof tape
    Prepared pulp
    Bone folder

Making the Cards
1. Cut a piece of deckle-making material to 8½" x 11" or the inner dimensions of your deckle.

2. Using a ruler, mark your cut lines according to the diagram at right. Cut out the four rectangular areas with a cutting knife, leaving a cross-shaped piece intact.

3. Fit the cross-shaped piece into the deckle, trimming off any excess if necessary, and tape it in place on both the top and bottom sides of the existing deckle, making sure that the deckle-making material is flush with the mould surface.

4. Place the new deckle on top of the mould, and make a sheet of paper. After the sheet drains, remove the deckle as usual, and couch the sheet.

5. When the cards are dry, use a bone folder and ruler to score the cards and fold them in half.

step 2

3³/₄" 1" 5" 1"

step 3

step 4

TIP

If you are creating a post of paper, it is a good idea to create a couching guide (see page 124) so that the sheets will be evenly aligned for pressing.

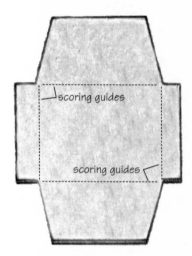

Envelope template — enlarge 400 percent or more.

step 2

Making the Envelopes

Follow the above procedure to create matching envelopes for your cards. When preparing the pulp, remember that you will need to address your envelope when it is dry and ready to mail — so the pulp should be neither too dark nor too textured (unless you plan to adhere a mailing label). When the paper is dry, assemble as follows:

You Will Need:
> Envelope template (at left)
> Card stock or cardboard
> Paper for envelopes
> Cutting knife
> Glue stick
> Wax paper (optional)
> "Relick" glue or double-stick tape
> Bone folder

1. Cut a piece of card stock or cardboard to the size of the dotted line on the enlarged envelope template marked "scoring guides."

2. Position the rectangular template on the outside of a dry envelope, and score around its edges with a bone folder.

3. Turn the envelope over, and crease flaps along the score lines. Fold the side flaps in toward the side that was not scored.

4. Carefully apply glue to the flaps. Fold the bottom flap up, and attach it to the side flaps.

5. You can purchase "relick" glue in some stationery supply stores or from paper-making suppliers (see resource guide). It can be brushed onto the top envelope flap and licked for sealing. Otherwise, you can just seal the envelope with a glue stick or double-stick tape when you are ready to mail it.

step 4

apply glue

MAKING A SHAPED PAPER BONBON BOX

Designer Julie Lau of Portland, Oregon, created unique sachets containing wildflower seeds as favors for her wedding guests. Friends from around the country received instructions in their individual packets to plant the seeds in honor of the married couple. The idea was such a hit that she decided to produce them for others.

Influenced by envelopes and packaging she has seen, Julie takes these ideas and adds her own twist. Origami has also inspires her innovations. Often she incorporates fabric, ribbons, and twigs into her paper creations. Her first design involved folding an 8½ inches x 11 inches sheet of paper to create a unique envelope that had ribbon closures. This has evolved into shaped sheets that fold up to create unique packages, such as the following bonbon box. (To purchase templates and instructions for the other designs, see resource guide.)

Note: This project demonstrates using a substitute deckle instead of a regular deckle.

You Will Need:

Deckle material
Mould and deckle
Photocopy of the bonbon template (at right)
Cutting knife
Prepared pulp
Glue (hot glue gun)
18" piece of ribbon (chiffon, silk grosgrain, or other)
Hole punch

1. Cut a piece of deckle-making material to the outer dimension of the mould.

2. Trace the outline of the template onto the deckle-making material, and cut out the shape. Make sure that the shape will fit inside the "image area" of the mould.

Paper can be shaped to create unique envelopes, favors, and other types of decorative packaging.

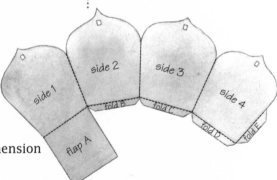

Bonbon template — enlarge 350 percent.

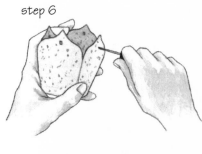

step 6

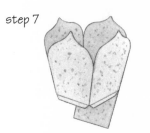

step 7

3. Place the new deckle on top of the mould, and make a sheet of paper. After the sheet drains, remove the deckle as usual, and couch the sheet.

4. After you have finished making paper, press and dry the sheets.

5. Using the template as a guide, crease the sheet along the dotted fold lines, and then lay the sheet flat.

6. Punch or cut holes in the four sides of the box as indicated on the template.

7. Fold sides 1 to 4 together to form the sides of the box.

8. Apply glue to folds B, C, and D, and fold flap A over them so that it covers the folds and is glued to them. Let glue dry.

9. Gently pull side 4 away from side 1, and apply glue to fold E. Tuck it in, and glue it to side 1.

10. Thread a piece of ribbon into side 1 and out of side 3.

11. Tie the two ends of the ribbon together, left over right as if you were tying the first part of a bow, tightening until the tips of sides 1 and 3 touch.

12. Pass one end of the ribbon through the hole in side 4 and the other through the hole in side 2.

13. Tie the ends of ribbon into a bow, pulling it tight so that the tips of sides 2 and 4 touch. Trim the edges of the bow.

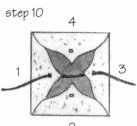

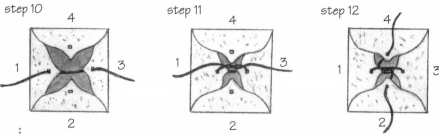

step 10

step 11

step 12

CHAPTER 11

WATERMARKING

A watermark is a translucent design incorporated into a sheet of paper when it is made; it becomes visible when the dry sheet is held up to light. Traditionally, watermarks were made by creating an image, such as a logo, in wire and sewing it onto the papermaker's mould. (This is still done with machine-made papers today, but watermarks are soldered to a wire roll for the commercial paper-making process). Since the wire is raised above the mould's surface, the paper made on the watermarked mould is actually thinner where the wire is and a regular thickness everywhere else. This variation in thickness makes the watermark more translucent than the rest of the sheet. After being pressed and dried, the variation in thickness is not apparent, but when the paper is held up to the light, the image becomes visible.

MAKING A WATERMARK

There are many creative ways to make watermarks, including the traditional method of sewing wire to the mould. When creating a watermark, consider how many times you will use it. If you want a logo on every sheet of paper you make, you can twist a wire design and sew or solder it to the screen. But once in place, it's difficult to remove! For a removable watermark, you can attach a design with a spray adhesive or make a separate screen to lay on top of the existing mould.

Watermarks are visible when the pressed, dry paper is held up to the light.

Suggested Watermarking Pulps

Pulp choice is important since watermarks become visible when held up to the light. Here are a few things to consider when selecting your pulp.

◆ Short hydrated fibers, such as cotton linters and cotton rag, work best for detailed designs (any of the techniques described later will work with these fibers).

◆ Long bast fibers work well with bold open line designs, but not with intricate designs (fabric paint and lace paper techniques work best with these fibers).

◆ Long, stringy, fast-draining pulps will not produce a well-defined watermark; the pulp needs to be consistent and smooth. Don't use irregular pulp or inclusions.

◆ If you're coloring the pulp, use only small amounts of pigment, because pigments opacify the pulp and make the finished sheets less translucent, obstructing the visibility of the watermark.

Note: Do not overshake when forming the sheets, as this will make it harder to obtain a crisp, clear image.

Watermarking Variations

Papermakers use a variety of methods to create watermarks. Wire, adhesive-backed materials, and fabric paint are all possibilities.

Papermaker's Project

WIRE WATERMARKS

You can design a watermark with wire by simply twisting a design and then sewing the wire to your mould's surface. A design composed of one line, rather than one that overlaps, will work best, since the overlapping will create joints that attract pulp. If you must have overlapping pieces, try hammering the joints flat — or you can solder individual pieces of wire together to create a flat design.

You Will Need:

16- or 18-gauge copper or galvanized steel wire (any nonrusting wire will work; don't use one that's too thick, because pulp might get caught underneath when you form sheets)
Fine wire, strong thread, or fishing line
Sewing needle to fit through the holes in the mould's screen
Needle-nose pliers

1. Twist the wire into your desired design, using a pair of needle-nose pliers if necessary.
2. Attach your watermark to the mould, making sure the wire does not interfere with the deckle. (I don't advise soldering the finished design to the mould, since it is difficult to remove, and you risk blocking the drainage holes in your screen). Use a thin, strong thread and a needle to sew the wire to the screen. Keep the stitches taut and close together so the wire will stay in place during sheet forming.

   Note: You can leave your design on the mould indefinitely, or simply snip the threads and remove it when you are done.

THE HISTORY OF WATERMARKS

When papers were first being created with watermarks, papermakers used the marks for different reasons: to signify when the paper was made, to designate the size of the sheet of paper, to advertise (as with a logo), or purely to add a decorative or symbolic element.

step 2

ADHESIVE-BACKED WATERMARKS

Since the principle of a watermark is to create a raised surface on the screen, contemporary papermakers have explored several alternatives to wire. One is to use a thin, flexible, adhesive-backed material, which is easy to cut and stick to the mould.

You Will Need:

Adhesive-backed materials (see box)
A simple line drawing or photocopied image (no larger than the dimensions of your mould)
Glue stick
Cutting knife
Spray adhesive (optional)
Mould and deckle
Prepared pulp

ADHESIVE-BACKED MATERIALS

+ Buttercut (a thin sheet of rubbery material approximately $\frac{1}{16}$" thick), available from Lee Scott McDonald
+ Sure Stamp, for making rubber stamps
+ Contact paper (an inexpensive alternative, but you'll have to stick several layers together to get a sheet thick enough), available in hardware stores
+ Craft magnet (a thin sheet of rubbery material intended for making magnets), sold in art and craft supply stores

1. Glue your drawing or image to the adhesive-backed material. This image will later be removed, and a glue stick doesn't seem to leave a residue. (You can also transfer your image using carbon paper or a graphite rubbing.)

2. Using a sharp blade, carefully cut out the image. Peel away the photocopy, and save it or throw it away.

step 2

3. Remove the adhesive backing from your design, and position the watermark on the dry mould. If the watermarking material is not self-adhesive, or if the adhesive is not strong enough to hold it on the mould surface, apply a thin coat of spray adhesive to the back side of the watermark. When the spray adhesive is tacky, press the watermark firmly onto the dry surface of the papermaking mould in the position where you want the watermark to appear. The spray adhesive may leave a residue on the mould's surface after you remove the watermark design. If so, clean the surface by rubbing it with a cloth dipped in acetone or lacquer thinner.

step 3

4. Now make a sheet of paper. Your sheet should be thick enough to at least cover the design you've attached with pulp. Once you have formed the sheet, you should see a faint outline of the attached design. If it is too thin, you will end up with holes in the paper. If it is too thick, you risk not being able to see the watermark in the sheet.

TIP

If you are using contact paper, cut several sheets slightly larger than your design. Next, strip off the backing sheets one at a time and laminate one sheet on top of the other. Carefully roll one sheet down on top of the next, to avoid creating air bubbles between the sheets. You will need 6 to 10 layers, depending on the thickness of the paper you're making.

FABRIC PAINTED WATERMARKS

This technique was developed by Mina Takahashi. Since wire and other stiff materials were not suitable watermarking materials for the flexible couching style of the *su* used in Japanese papermaking, Mina searched for a flexible material to create a raised watermarking surface. Fabric paint is both flexible and permanent, and it can be squirted onto a *su*. You can make multiple reusable watermarking surfaces using this technique. It can be used in Western-style paper-making too, but it is not suitable for the deckle box technique.

These instructions assume that you will use a template, but you can just "draw" your watermark freehand if you like.

You Will Need:

Mould and deckle or *sugeta*

Fiberglass window screen or no-see-um netting (avoid mosquito netting, which is not stiff enough)

A drawing or photocopied image (no larger than the dimensions of your mould)

Transparent Mylar, Plexiglas, or plastic sheeting

Fabric paint (use a brand that's a slick plastic paint, not the type that has to be ironed to puff up; I buy Tulip brand)

Prepared pulp

step 3

1. Cut a piece of fiberglass window screen or no-see-um netting to the outside dimensions of the mould.

2. Place your drawing or image on a working surface, and cover it with a piece of Mylar. Put the fiberglass screen or no-see-um netting on top, and tape the stack in place. Stretch the screen or netting so that it is as taut and flat as possible.

3. Using fabric paint, trace the image onto the screen. (You might want to practice squirting the fabric paint on a scrap of paper first, to get a feel for drawing with it.)

4. Leave everything in place until it dries (at least overnight). When ready, the fabric paint should feel dry and firm to the touch. If you lift it too soon, some of the fabric paint will seep onto the Mylar and damage your design.

5. Check the evenness of your drawing. The fabric paint should be about ¹⁄₁₆" thick in all areas. If you like, you can touch up thin areas and let it dry again.

6. To make paper, sandwich the screen or netting between your mould and deckle — or in your *sugeta* — and make a sheet. For Western papermaking, the screen will couch with your sheets, and you will need to peel it up and place it back on the mould. Or, you can tape the flexible screen to the mould, and you might need to sew it down at a few points in the middle. If you use the adapted Japanese papermaking technique, this screen is your adapted *su*. If you are using a traditional *sugeta,* you will need to sew the no-see-um netting onto a bamboo mat for support.

step 6

PAPERMAKER'S PROJECT

PAPERMAKER'S TEARS

Water droplets that hit a freshly formed sheet of paper will actually displace the pulp on the sheet and create a slight indentation, resembling a watermark. Although these are usually mistakes (they are generally referred to as papermaker's tears) this technique could be used to create decorative sheets of paper.

You Will Need:
> Mould and deckle
> Prepared pulp
> Water
> Spoon or other spreading tool
> Eyedropper or turkey baster

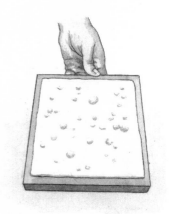

Water droplets that hit fresh paper are called "papermaker's tears."

1. Form a sheet of paper, and rest the mould (with the fresh sheet) face up on a flat surface. Dip a spoon or other tool in water, and sweep it across the mould, dripping onto the sheet. Or, if you want to create a more regular pattern, use an eyedropper or turkey baster to create the drips.

2. Couch the sheet.

MAKING LACE PAPER (A PSEUDO-WATERMARK)

Papermaker Mina Takahashi showed me this traditional Japanese technique, which can be used to create papers that look just like lace. In Japanese, this paper is called *moyogami*.

Note: You will need a long fiber for this, because you will be spraying the wet sheet of paper and breaking bonds, but the fibers still need to hold together. Suitable fibers include abaca, kozo, mitsumata, and gampi. (You can use shorter fibers, such as linen and cotton, to create lace shapes, but they won't hold together as entire sheets and will need to be laminated to a base sheet.)

You Will Need:

>   Old picture frame (or extra deckle)
>   Lace stencil materials (see box on page 159)
>   Duct tape
>   Mould and deckle
>   Prepared pulp
>   Garden hose or watering can (the key here is sufficient water pressure, so you'll need a spray nozzle on your hose or, if using a watering can, you will need to stand on a ladder)

1. Use duct tape to attach the lace to an old frame or extra deckle. Stretch the lace as tight as possible, because it will bow when it gets wet. The frame should keep the lace at least ½" above the surface of

VARIATION

You can use a dental syringe (available from some papermaking suppliers or medical/dental supply companies) or any other device that squirts a steady stream of water to create imagery or patterns by displacing the fibers on the mould. Squirt onto a freshly formed sheet of paper that is still sitting on the mould surface and has not yet been couched.

the paper, because the lace will bow and should not touch the paper. If your lace and frame don't need to be used for anything else, you can use waterproof staples (Monel brand) instead of duct tape, but do not staple into the deckle. If you are using metal grating, you can rest it on the deckle or another support so that it almost touches the paper, since it won't bow.

2. Make a sheet of paper, and set it on the mould on the floor (it must be raised off the floor by at least 1" so that it can drain properly. If you are forming your sheets on a *su,* place the *su* on top of the Western mould so that you have approximately 1" of drainage space beneath it.

3. If using cloth lace, mist or dampen it. Set the framed lace or stencil above the sheet of paper by resting it on something that raises it about 1" above your wet sheet — a second deckle will do the trick.

step 2

LACE STENCIL OPTIONS

✦ Decorative aluminum metal sheeting used to cover radiators, available at building supply stores (good because they stay flat and don't wear out)

✦ Lace with openwork (⅛" holes or bigger) (I don't recommend using heirloom laces)

✦ Plastic lace by the yard or roll, available at dime stores (can be used one time only because they tend to rip when tape is removed, or, you could leave them on the frame permanently and reuse them)

step 4

step 5

4. Apply an even flow of water pressure from approximately 3' above the sheet. Test the sprayer or watering can first to make sure you can get a consistent spray of water. Spray back and forth evenly over the surface of the lace, keeping the spray perpendicular to the lace. For a thin sheet, spray over the surface for about 10 seconds.

5. Remove the lace, and couch the sheet of paper.

Tip: You can couch the lace paper onto a backing sheet with items such as leaves or feathers embedded between. The holes in the lace will create "windows," making the embedded objects more visible.

COLOR YOUR MARK

Since watermarks are generally not visible unless held up to a light, try couching a thin, watermarked sheet, made with a white or light-colored pulp, onto a darker pigmented sheet that isn't watermarked. The darker color will show through the watermarked area, making the design visible on the flat sheet.

CHAPTER 12

CASTING

You can press freshly made sheets of paper or pulp into low-relief molds to create detailed lightweight paper castings. Cast directly with paper pulp, or work with sheets of paper that have been formed and lightly pressed. There are many techniques you can use to create the object to be cast, or you can simply cast directly onto an existing object, such as a bowl. (In the latter case, you need to use an object that will not be damaged by wet pulp.)

Two common mold-making methods use plaster molds and rubber molds. With each method, you first create a "positive" — an object that looks exactly the way you want your casting to look (usually in clay) — from which you will make a plaster or rubber mold. You then use the plaster or rubber "negative" to recast your original piece in paper.

The techniques described in this chapter are suitable for casting images up to 3 inches deep. To create larger three-dimensional sculptures, I recommend making an armature (see chapter 13) or casting in parts.

CHOOSING YOUR PULP

In general, it is best to cast with a low-shrinkage pulp — such as cotton, recycled paper, lightly beaten abaca, or a suitable plant fiber — so that the cast paper will not shrink out of the mold as it

Casting is a way to create three-dimensional objects out of handmade paper.

161

Working in layers will make your casting strong enough to be free standing.

dries. When casting into a concave mold, a fiber that shrinks too much will pop out of the mold, and details will be lost. When casting onto a convex mold, the fiber will shrink to the form and be difficult to remove.

If you want your casting to be strong enough to be free standing when dry, you should cast with a thick pulp rather than a thin one. You can cast in layers to build up the thickness, or, if you have a limited quantity of a particular pulp, you can cast a thin outer layer with that pulp, and then cast another layer of thicker pulp behind it to add strength and support.

DEFYING NATURE

Although low-shrinkage pulps work best for casting, artist Mary Ting wanted to use overbeaten abaca, a high-shrinkage pulp, to make her cast pieces translucent. She created forms by sewing fabric together in the shape of long cones, which she then filled with sand to make them three-dimensional and rigid for casting. She hung the forms from the ceiling and cast over them with freshly made sheets of abaca. As the pulp dried, it did shrink to the forms, but she was able to simply pour out the sand, collapse the bags, and remove them when the paper was dry. This is an excellent solution for casting with a high-shrinkage fiber.

Artist Winifred Lutz has also experimented widely with creating unique molds for casting with high-shrinkage pulps. Her article "Casting to Acknowledge the Nature of Paper" appears in *Hand Papermaking* (Vol. 1, No. 1, spring 1986).

CREATING A MOLD

Nonporous materials such as clay work best for creating the original from which to make a mold for casting. If you use a porous material, such as cardboard, the plaster or rubber molding compounds will penetrate into the pores, stick to the material, and damage it during contact. It is possible to use some porous materials only once. For example, you could pour plaster or rubber onto a cardboard mold original, ripping the cardboard out of the mold when it dries. (You may also have to scrub bits of cardboard out of the mold.) For repeated use, you can seal porous materials, thus making them non-porous, with a sealant such as polyurethane. This will result in an original that will withstand the mold-making process; just remember that the polyurethane is permanent, so don't coat a treasured heirloom with it!

A common material used to make originals is plasticene clay, an oil-based product found in art supply stores. This clay does not dry, so you can reuse it if you take care to remove the bits of plaster that may stick to it during mold making — or you can discard it. You can hand-form or use tools to carve your image in clay. Create a "positive" of the image in clay, which looks exactly like the end product you desire. Make the positive in low relief — a flat surface that is no deeper or taller than 3 inches — and take care not to create undercuts.

Avoiding Undercuts

When making a mold, you need to make sure there are no edges protruding from the positive that would prevent it from being separated from the mold once it is dry. (These edges are known as undercuts.) For example, if you were to cast a solid trapezoid block with the heavy side down, it would be impossible to remove from the mold when dry. However, if you flipped it over with the heavy side up, it would slip right out — in this case, there would be no undercuts.

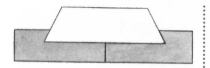

Cast an object with unavoidable undercuts in two or more sections, then join the two sections when dry.

One way to cast an object with unavoidable undercuts is to cast it in two or more sections, then join the casts of the sections when they are dry. Right angles are also difficult to remove from the mold, so if possible, taper the form slightly to aid in removal from the mold.

The right way to cast an object to avoid undercuts.

The wrong way to cast an object. The wider edges (undercuts) below the surface make it impossible to remove when dry.

Creating a Low-Relief Plaque

If you are creating a mold from found objects, attach the objects permanently with glue or temporarily with clay onto a sheet of Plexiglas or glass, creating a low-relief image. For example, you could attach shapes cut out of Plexiglas. You could also lay out a slab of plasticene clay and embed your objects in it. You can even use cardboard, but your molding compound will most likely stick to the cardboard a bit and will take some care to remove when dry, unless you seal it with coats of polyurethane, acrylic spray varnish, or wax.

Regardless of the materials you use to create the positive, build a box around the plaque with pieces of Plexiglas, foam core, cardboard, or wood. Seal the edges with a waterproof tape or clay so the molding compound does not seep out. Or use a cardboard box lined with a plastic bag, but make sure the item you are casting is attached to the bottom, so that the rubber or plaster doesn't slip underneath it.

Seal the edges of your mold with waterproof tape or clay.

MAKING A RUBBER MOLD

This is a two-part rubber molding compound that comes with mixing and setting instructions. It has an indefinite shelf life until it is mixed. Once mixed, it hardens overnight. Twinrocker also carries a rubber molding compound and casting supplies (see resource guide).

You Will Need:
> Rubber molding compound (PMC 724 from the Smooth-on Company works well for paper casting. Smooth-on also makes other compounds with varying degrees of hardness.)
> Casting positive

1. Follow the directions on the molding compound package, mixing enough rubber to fill the bottom surface of your "box."
2. Pour the mixed rubber into the box, making sure it covers the highest point of the original by at least ⅛" (for one-time use) and ½" (if you plan to make multiple castings from the mold). You can mark a fill line on the side of the box so you know how much rubber to pour.

MAKING A PLASTER MOLD

These molds use pottery plasters, which are more absorbent and stronger than plaster of Paris.

You Will Need:
> Disposable paint bucket or other plastic bucket
> Plaster (Hydrocal, U.S. Gypsum Molding Plaster #1, and Georgia Pacific K 59 are three good brands available at ceramic supply stores)
> Stick for stirring
> Casting positive

RUBBER OR PLASTER?

Either type of mold works well, but rubber and plaster have their own distinct advantages.

Rubber
+ More durable for multiple casting
+ Can be put in a press
+ Flexible and easier to release from

Plaster
+ Less expensive
+ Easier to find
+ Absorbs water and dries more quickly

step 1

step 3

1. Fill your bucket with an amount of cold water that approximates the volume of plaster you need. Add a handful of plaster, sprinkling it over the entire surface of the water. Continue to sprinkle plaster over the surface by the handful until the plaster starts to float in islands on the water's surface (before it sinks).

2. When the islands float for approximately 3 seconds before they sink, stop adding plaster, and stir the mixture thoroughly with a stick, taking care to mix in as little air as possible. As the plaster gets warmer you'll notice it starting to thicken.

3. Pour the thickened plaster into the mold, making sure that it covers the original by ⅛" to ½". Don't wait too long — if the plaster gets too hard, it won't flow well, and your resulting mold will be brittle and powdery. On the other hand, if you don't have enough plaster, the mold will be soggy and crumbly.

4. There will be a residue of plaster left in the bucket. You can either discard the bucket or allow the plaster to dry, then break it into pieces with a hammer and remove it.

MOLD-MAKING TIPS

✦ Remember that you will need enough rubber or plaster to cover the original or positive by at least ⅛". If you will be using the cast repeatedly, cover your original or positive with at least ½" of rubber or plaster so that it will endure multiple castings. The entire depth of the rubber should be no more than 3", to make it easier to release the objects from the mold.

✦ Make sure that the tabletop or surface the mold box rests on is level while you're pouring and as the mould dries.

✦ Both plaster and rubber will dry to the touch in a couple of hours, but you should let them dry overnight before trying to remove the original from the mold.

CASTING RUBBER OR PLASTER MOLD WITH PAPER

There are two methods of casting. You can 1) form sheets of paper, press them lightly, and then cast the wet sheets (laminate casting), or 2) cast directly with wet pulp. If done properly, your casting will pick up an incredible amount of detail and will really resemble your "positive."

LAMINATE CASTING

The advantage to laminate casting is that the fibers bond better than during casting with loose pulp, resulting in a casting that is more even. And since the sheets are drained and lightly pressed, you will have better control over the thickness of the piece, and it will dry more quickly with less sponging required.

You Will Need:

Mould and deckle
Prepared pulp
Plaster or rubber mold
Brush
Sponge
Interfacing or cloth
Methyl cellulose
Fan (optional)

1. Make several sheets of paper, and either sponge the sheets one at a time or press a post of sheets to remove excess water. If pressing, do not press fully. The sheets should retain some water, but you need to be able to handle them without pulling them apart. Do not make the sheets too thin, or the resulting casting will not be strong enough to hold its shape.

USING MOLD RELEASES

A release agent applied to the surfaces of the rubber or plaster mold will allow you to separate the casting when it is dry. Release agents include:

◆ Parfilm (made for casting; no residue, low odor, no mess)

◆ Silicone spray (available in hardware stores)

◆ Vegetable oil cooking spray

◆ Green soap (made for casting, requires cooking)

◆ If you are using a silicone spray, apply it outside or in a well-ventilated area.

2. Tear off a 4" square piece of the wet paper (you may need to adjust the size depending on the intricacy of your mold). Drape the torn piece of paper into the plaster or rubber mold, and use your fingers, a brush, or a sponge to press the paper into any crevices.

3. Drape a small piece of interfacing or cloth (to protect the paper from the sponge) over the wet paper, and gently sponge out some of the excess water.

4. Tear another piece of paper, and overlap the piece you have just laid down, laminating the two pieces together by pressing again with the sponge. These feathery, torn edges will be less visible than hard, cut edges. In addition, if you keep the edges of the sheets wet, they will blend, and the overlap won't be as likely to show when dry. I also recommend brushing a bit of methyl cellulose (see recipe on page 75) onto the seams, to ensure bonding.

5. Repeat this process until you have covered the entire surface of the mold.

6. Leave the mold in a well-ventilated area to dry. Once thoroughly dry, remove the casting from the mold.

step 4

step 6

CASTING WITH LOOSE PULP

The advantage to using loose pulp is that it's easier to work the pulp into small grooves. And if the shape of your mold might cause the formed sheet to break, you should definitely opt for using packed pulp.

You Will Need:
Prepared pulp
Plaster or rubber mold
Interfacing
Sponge

step 1

1. Place a small amount of pulp onto the object and press it into the plaster or rubber mold, using a sponge. (Place a small piece of interfacing between the pulp, and sponge when you press.) The layer of pulp should be approximately ¼" thick. As you remove water with the sponge, the pulp will be compacted and become thinner; when it is dry it will be even thinner, so start with a fairly thick layer.

2. Keep filling in the mold with small amounts of pulp, making sure you sponge very well at any seams to ensure bonding.

3. After you've filled the entire piece with pulp, go over it again with the sponge to remove as much water as possible.

TIP

You can add extra sizing to make the casting dry harder. Sizing also has a gluelike quality that will strengthen the fiber bonds. You can also add methyl cellulose or PVA glue (approximately 2 tablespoons per gallon of pulp that is the consistency of cottage cheese). If you will paint the casting later, make a small test piece first — sizing and glues will make the casting less absorbent.

PRESSING ALTERNATIVES

If your cast piece is in low relief, it is possible to press the water out in a hydraulic press or by using a vacuum table. I do not recommend putting anything breakable in a hydraulic press, because you'll risk cracking it. You might need to test how much pressure you can apply without deforming the mold. When pressing, just make sure to pack the press with extra felts and/or foam on both sides of the piece to compensate for both the thickness and the relief areas. If you are casting multiples from the same mold, you might want to cut the packing material to fit into the hollow of the mold for efficient pressing.

DRYING SUGGESTIONS

It is easier to make a hollow casting rather than fill the mold with solid pulp. Not only would a solid casting take longer to dry, but it would not dry effectively. To make a full-round casting, it is better to create two halves and then seam them together when they are dry.

Use a fan to help circulate air in the room where you dry your cast paper objects. Don't just point it directly at the casting, because it might cause it to dry unevenly and pop out of the mold. To achieve the finest detail, let the piece dry in the mold. If you remove it before it is dry, you'll lose detail, and the piece will warp.

You can weight down flat parts of your casting with bricks or other heavy items to restrain it as it dries.

Heavy objects can be used to weight down the flat parts of your casting as it dries.

CASTING A BOWL

This project requires a wet/dry vacuum cleaner to draw pulp onto a shaped mesh form. A unique method of casting, it was developed by Lee Scott McDonald of Charlestown, Massachusetts, who designs and constructs custom and production papermaking equipment in addition to running a mail-order papermaking supply company (see resource guide).

You Will Need:

> Kitchen strainer (or any wire mesh sieve with or without handles; the mesh should be at least 20 lines per inch and fairly strong — it should not dent with force, or it will collapse during this technique)
>
> Piece of ½" plywood or plastic board, large enough to cover the opening of the strainer
>
> Hole saw or jigsaw
>
> Wet/dry vacuum cleaner with hose
>
> PVC pipe 2" in diameter (optional)
>
> Waterproof tape or rubber gasket
>
> Vat of prepared pulp, at least 2" deeper than the strainer
>
> Stick for stirring
>
> Pulp (any short fiber, such as cotton or recycled pulp; longer fibers tend to become stringy)

1. Cut a piece of plywood or plastic board to fit the strainer. You can leave it square, but it will be more difficult to attach to the strainer.
2. Cut a hole in the plywood or plastic board, large enough to feed the vacuum hose all the way through (you can feed a piece of 2" diameter PVC pipe through the hole instead and attach the vacuum hose to that).

3. Tape the strainer to the board, making an airtight seal. (You can try taping rubber gasket around the edges instead — if the vacuum is strong enough, the suction will hold the strainer in place during formation.)

4. Insert the hose or PVC pipe through the hole so that it comes within ¼" of the strainer. You may have to adjust this distance to get the best results. Tape the hose or PVC pipe in place, making an airtight seal. It is better to use several small pieces of tape and do a neat job than to wrap a long piece around the hose or pipe. If you use pipe, you can also use caulk or plumber's putty to make the seal.

5. Make sure the vacuum cleaner is ready for wet work (remove the cloth bag if it has one). Plug it into an outlet (one with a ground fault interrupter [GFI] is best). Always use caution when working with water and electricity. Hook the hose onto the vacuum cleaner.

6. Fill the vat with pulp and water. Practice will tell you how much pulp to charge the vat with. In general, you will get finer detail if you keep the vat thin and leave the vacuum contraption in longer. Stir the fiber in the vat with a stick (or use your hands).

7. Turn on the vacuum cleaner, and insert the entire contraption into the vat, bowl-first. Count to five, then remove the contraption from the vat. If the strainer is not covered with pulp, resubmerge it.

8. Remove the strainer from the board, and let the piece air-dry. When dry, pop the bowl off the strainer.

CHAPTER 13

USING ARMATURES

An armature is a support structure for a sculpted piece. You can use many things to create armatures for paper, including wire, reeds, wood, and branches. This involves constructing the armature first and then using this structure as a base to wrap your damp paper over to create sculptural forms that can be air-dried.

In addition to creating armatures, you can work on found objects, such as old lampshade frames, garden trellises, or baskets. Many times, items that make very good armatures can be found at garage sales, in consignment shops, or even in your own attic — and find new life and purpose with a creative wrapping of handmade paper.

ARMATURE TECHNIQUES

When covering your armature, you can make shaped sheets to cover certain areas (this technique is best if you have several areas that are the same shape, such as the panels of a lampshade), or you can make full-sized sheets, which you can tear or cut to the desired shape and size and attach them to the armature in a manner somewhat similar to the laminate casting technique described in chapter 12. Several types of armatures are described in the chapter on pulp spraying, and many of them would be suitable for covering with sheets of paper as well. You can choose to remove your armature after covering or keep it as a structural element of your piece.

Selecting Pulp

Any pulp is suitable for wrapping over an armature, but you should consider translucency and shrinkage. If you want translucency, use an overbeaten fiber. If the armature is a soft form, a high-shrinkage pulp could crush it, or it could create a new and interesting form.

Other considerations:

◆ Polyurethane wood parts so that they are waterproof
◆ Use galvanized metals (or treat other metals so they don't rust — unless you want them to)
◆ Decide whether or not the armature will be removed (this will determine how you construct your piece)
◆ Use an adhesive to adhere the paper to the armature frame (I recommend PVA glue)

PAPERMAKER'S PROJECT

PAPER LANTERN

Judy Hoffman, a New York–based paper artist, showed me how to make this lantern, which traditionally features a wooden support structure. She learned the process at the Paper and Book Intensive, an annual gathering of book and paper artists. Timothy Barrett and Richard Flavin, who are both skilled in Japanese papermaking, taught the course, which was based on traditional Japanese lantern making.

You Will Need:

Pencil
Card stock
⅛" foam core
Cutting knife with sharp blade
Straight pins

Transparent tape

Thin basketry reed

7 strips of Japanese tissue or other strong paper cut
    to ½" x 2"

Glue brush

PVA glue

Small clips or clamps

Handmade paper (you can create shaped sheets to cover
    the form, or you can work with premade sheets)

Making the Lamp Frame

1. Enlarge the patterns at the right by 400 percent. Cut out the lamp rib and spoke patterns, and with a pencil trace them onto card stock to make a template. Cut out the templates.

2. Trace two spokes and eight ribs onto a piece of foam core, and cut them out with a cutting knife. (Make sure your blade is sharp — a dull blade will make cutting more difficult and slower.) Trace the marks on the rib pattern onto the edges of each rib to indicate the reed spacing.

3. Assemble the lamp form so that it resembles the diagram. If any pieces are loose, stick straight pins into the foam core to hold things in place.

4. Apply strips of transparent tape to the edges of the ribs, to keep the paper from sticking to them when gluing.

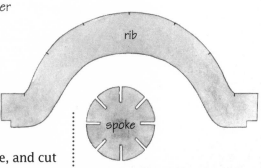

Lamp frame template — enlarge 400 percent.

rib

spoke

step 2

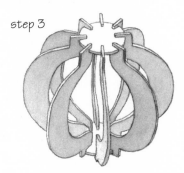

step 3

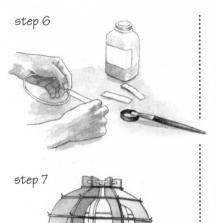

step 6

step 7

step 1

5. Take a piece of reed and wrap it around the top section of the lamp frame, where it starts to curve. Cut the reed so it's as long as the lamp's circumference plus ½". Wrap the reed around the lamp frame again, and mark where the ends overlap. With a cutting knife, carefully slice the two ends of reed at an angle, so they will have a flush joint.

6. With a brush, apply PVA glue to one side of a strip of Japanese tissue, and wrap it around the reed joint. Place the reed ring on the lamp frame. You might try clamping the ends of reed together with a small clip or clamp until the glue sets.

7. Repeat steps 5 and 6, measuring pieces of reed to fit around the ribs at each subsequent marking. Use straight pins to hold the rings of reed in place on the ribbed frame.

Applying the Paper

1. Cover the armature with paper (cut out from the template below) as if you were covering sections of a grapefruit: Remove the straight pins from one section, and apply glue to the sections of reed. Place one piece of paper over that section, centering it over the two ribs. As you work, you will need to manipulate the paper a bit to compensate for the curvature of the armature. You might even need to make tiny pleats or folds. Gently misting dry paper before applying it will make it easier to manipulate, too.

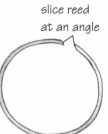

slice reed at an angle

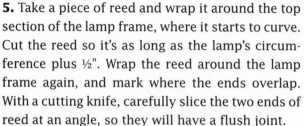

Paper template (to cover lamp frame) — enlarge 400 percent

2. Remove the straight pins from the section directly across from the first paper section, and apply glue to the reed sections. Apply a sheet of paper as in step 1.

3. Apply paper sections to every other section until you have covered four sections. Applying the paper sections in this fashion holds the rings in place.

4. Next, cover the remaining sections. In addition to applying glue to the reed strips, apply a ¼" strip of glue along the two overlapping sides of the paper panel. Place the panel on the armature, overlapping the paper on both sides and tamping it down with your fingers or a brush to make sure it adheres to the adjacent sheet. Repeat until you have covered every section.

5. Allow the shade to dry overnight.

6. Remove the armature by simply pulling out the spokes, collapsing the ribs, and gently pulling them out of the paper form.

step 6

7. If you plan to hang the lantern, tie a piece of string about 10" long to the top reed ring, in two places directly across from each other. Apply glue to the extra ½" of all sections at the top and bottom of the section of paper, wrapping it around the reed as you go, and gluing the paper to itself. Cut little slits when you work around the string.

Variations: You can make these lanterns in many shapes and forms. When designing new shapes, just make sure that the form can be removed from the opening in the lamp after the paper is applied. To vary the paper, you might collage sheets of paper together to make the paper panels, or cut the edges of the paper pieces using deckle-edged scissors to make a decorative overlap.

PAPER LAMP COVERINGS

Try a variety of papers to cover the lantern:

◆ Use the paper pattern on page 176 to make a shaped deckle, and make eight shaped panels of paper (if you make paper, press sheets but don't dry them).

◆ Use the pattern to cut eight pieces from a thin and flexible dry sheet of paper (most Japanese papers work well).

◆ Use assorted damp handmade papers or dry papers cut or torn into various sizes and shapes, and cover the form in a random fashion. If you use this technique, apply methyl cellulose glue where the paper overlaps.

FRAMED PAPER PANELS

You can "stretch" wet handmade paper over a simple wooden frame, simulating a canvas stretcher. When dry, your piece will be ready to hang. The following instructions are for an 8" x 10" frame; if you plan to make a larger frame, you should brace the corners.

You Will Need:

2   1" x 2" x 9½" pieces of wood with mitered corners
2   1" x 2" x 11½" pieces of wood with mitered corners
        (smaller widths of wood will warp; larger is okay)
Galvanized finishing nails or brass screws
Drill, hammer, screwdriver
Polyurethane or other sealant
Brush
Mould and deckle
Prepared pulp (at least some shrinkage content in the pulp will
        make for a stronger canvas. I recommend 25 percent
        abaca, 75 percent cotton — this will also make the paper
        shrink and fit the frame well)

step 1

step 5

1. Join the 2" sides of the 1 x 2s with nails or screws to make a frame.
2. Using a brush, seal the wooden frame with the polyurethane.
3. Make a sheet of paper that is approximately 15" x 17". The sheet is large enough so that it will stretch around the frame, but keep in mind that the image area is 8" x 10". You may want to extend the image around the edges of the frame to give it a cohesive appearance.
4. Press the sheet, and lay it face down on a felt.
5. Center the frame on the pressed sheet, and draw the borders of damp paper up over the back of the frame. Use a sponge to press the overlapping edges together to create seams.
6. Lift the framed piece off the felt, and set it face down on a dry felt to dry.

CHAPTER 14

PULP SPRAYING

Spraying with pulp is a unique way to form sheets of paper or sculptural objects. After beating the pulp, place it in a pulp sprayer — a pneumatic spray gun attached to an air compressor — and aim it at anything you wish to coat or encase with pulp. Many artists have used this technique to create a variety of sprayed papers and forms. Artist Joyce Schutter of Iowa wrote her thesis on pulp spraying when studying at the University of Iowa; the information in this chapter is derived from Schutter's work in the field.

NECESSARY EQUIPMENT

To spray with pulp, you will need a gun-type pneumatic pulp sprayer and an air compressor. The sprayer, which has a plastic hopper that gets filled with pulp, has several adjustable orifices, a wing-nut stop to set the flow rate on the trigger (if desired), and a notch to set the trigger for continuous flow (great for large jobs). Some models come with an attached shutoff valve to regulate the flow of air, which is well worth the added investment.

This type of pneumatic sprayer was originally designed for spraying plaster or concrete onto ceilings and walls, and is often stocked at rental companies (look in the yellow pages of your phone book).

To maintain your pulp sprayer, remove any pulp residue — from both the gun and the hopper — with water after each use. Also oil the

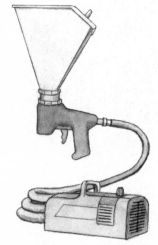

A pulp sprayer allows you to cover large areas with pulp, opening up many options for creativity.

sprayer after each use with pneumatic tool oil, applying a few drops to the air stem and the spring. Apply more oil if the trigger sticks.

You will also need an air compressor (available at hardware and building supply stores) with a minimum of 1.5 to 2.0 horsepower to operate the sprayer. (Schutter uses a 4 horsepower portable model.) The compressor must be able to maintain a minimum of 60 pounds of pressure per square inch (PSI) for smooth operation. If you use a small compressor (1.5 horsepower) you will need a holding tank to store the compressed air. Be sure to follow the directions and cautions that come with the equipment. Compressors are also available at rental companies.

PULP CONSIDERATIONS

You can spray many types of pulps, but abaca, cotton half-stuff (a type of cotton fiber), and raw and cooked flax processed in a Hollander beater tend to work best. When preparing pulp, think about the texture and opacity you want in your final piece. Schutter's favorite pulp is overbeaten raw flax. Its consistency is like that of overcooked or finely strained applesauce, and it requires no formation aid. It is highly translucent, and though it shrinks a lot with heavy applications, it does not shrink when applied thinly.)

Long fibers, particularly those used in Japanese papermaking, tend to be problematic. The longer the fiber, the more likely it is to clog the sprayer, which can be frustrating. When the sprayer clogs, the air flow continues, but the pulp stops flowing. Two things can happen: if you are spraying close to the work, the air can displace the pulp you've already sprayed onto the surface; or, the clog may dislodge a large, stringy clump of pulp into the middle of your work. At best, you will frequently have to stop and clean clogs, which can interrupt the consistency of your layering.

Using Formation Aid

After preparing the pulp, test it in the pulp sprayer. If it clogs, add formation aid (see recipe on page 74) in small increments, testing it each time you add an increment. (When working with cotton rag half-stuff that has been beaten for 3 hours, Joyce adds 1 cup of formation aid per 4 gallons of prepared pulp — right out of the beater, not strained or diluted.) Formation aid facilitates the flow of pulp through the sprayer, making it possible to spray longer fibers, but the addition of formation aid and water can also complicate the layering process by adding to the spraying time and to the number of layers needed to reach your desired thickness. Since the pulp is so diluted, it also increases the likelihood of pulp slides (areas in which pulp hasn't drained enough and starts to slide down the surface of the spraying area). This is especially true on three-dimensional objects.

Adding Stiffeners

You can add sodium carboxymethylcellulose (CMC) to your pulp as an internal stiffener. Besides strengthening thin layers of paper, it enhances their translucency. Schutter recommends 4 to 8 tablespoons per 10½ ounces of dry flax, added in the final few minutes of beating. She doesn't use it with opaque pulps because they are not visually enhanced by CMC. You can also use CMC as an adhesive, to join sections of a large piece, or as a brushed-on solution to stiffen a piece, applied at the wet or dry stage while the piece is on its armature.

Accounting for Pulp Shrinkage

As a general rule, highly beaten fibers tend to be more translucent and crisp and to have a higher shrinkage rate (as much as 45 percent). The thicker the application of pulp, the more pronounced the lateral (side-to-side) shrinkage will be. Shrinkage is also influenced by the surface resistance of the material it is sprayed onto. For example, a layer sprayed onto plastic will shrink more than one sprayed onto cloth, even if both substrates are fastened to a stable armature, because the plastic has less surface resistance. This will not be the case if your armature is completely three-dimensional, because the shrinking paper will envelop the armature, eliminating the possibility of popping off. However, if there is enough surface tension, the paper itself could split open.

You might want to make a test piece to see how the pulp will shrink as it dries. Be sure to make it large enough and similar enough to your desired final form so that the results will be accurate.

Planning Ahead

It's a good idea to figure out how much pulp you will need for a particular project. Make a test piece by spraying the pulp you will use in the desired thickness on a similar armature or screening material. Cut the dried test piece to a standard size (1 square foot, for instance). Remove the dried piece from the armature, and weigh it on a kitchen or postal scale. If you will not be removing the dry piece from the armature, you need to weigh the armature before spraying, then weigh the armature with the sprayed pulp and subtract the weight of the armature to get the weight of the paper.

Next, calculate the total surface area of your desired piece in square feet, and multiply that by the weight of 1 square foot. For example, if 1 square foot of paper weighs ¼ pound, and the total surface area of your piece is 8 square feet, you will need to prepare 2 pounds of pulp (dry weight).

This is well worth the time, especially on a large piece with highly beaten pulp. If you run out of a pulp that requires 8 hours to beat and have been applying layers every hour, keeping the piece wet enough to take the next layer from the new load can be difficult, especially if it is a large, vertical piece.

SETTING UP TO SPRAY

Your armature must be anchored securely so that you can turn it or move around it, and so that it will not be knocked over by the force of spraying it with pulp. In addition, try to anchor it so that you will be able to spray the entire piece without having to touch it.

In many cases, you can suspend your armature from above. If possible, it is also a good idea to anchor it to the floor beneath the piece. Swivel hooks anchored to the ceiling and floor work well; if you aim the sprayer right, the air can gently spin your mould so that you don't have to touch it at all. This works best with round or cylindrical forms.

If you are spraying on only one side of a piece, you can mount it on a frame or on sawhorses. How you mount your piece will depend on its size, shape, and weight. If a piece is heavy, you may not need to attach it at all. However, if it is light, it must be anchored. Keep in mind that the water must be able to drain freely beneath the piece.

Suspending your armature from above means you can spray the entire piece without touching it.

Protecting Your Work Area

Protect your work area by covering it with plastic sheeting — the walls, the floor, and even the ceiling if necessary. As you spray, quite a bit of pulp will go beyond your piece. It often isn't visible when wet, but it can really show up when dry. (This overspray can be quite beautiful — you can peel it off the plastic sheeting to get a paper with a lacelike quality.)

Water will collect both in your work area and around the armature and its supports, so make sure you devise a drainage system. Here are some things to think about. If you have a wooden support for your armature, it will hold water underneath your screen or fabric, so you must cut holes or slits in the wood and tilt it to facilitate drainage. Wood supports or frames will warp if left wet for long, so seal wooden parts. Plastic is slippery when wet, so figure out a way to walk around your piece safely while working.

Use a sawhorse mount for spraying pieces on only one side.

The overspray from a pulp sprayer has a beautiful lacelike quality.

Creating a drainage system beneath your piece allows you to recycle oversprayed pulp.

Here's a simple way to form a drainage system directly beneath your piece: You can create an inverted plastic tent in a funnel form that drains into a basin or bucket directly beneath a piece. This can also help you recycle the oversprayed pulp that collects.

Protect yourself, too, by wearing a waterproof apron and rubber boots. Some people are sensitive to airborne fibers and begin coughing in a room with pulp spraying. Wear a mask if you need to.

Operating the Pulp Sprayer

Before you start, establish a resting place for your pulp sprayer. You may need to set it down quickly to do something else, and you risk spilling the contents of the hopper because of its topsy-turvy shape. Setting it in a bucket works well.

1. Adjust the orifice size: Loosen the orifice tension nut and the orifice hex nut with a socket wrench (pliers tend to mar or strip the nuts). Pull the trigger back at least ⅛", and rotate the orifice plate to the desired size, aligning the hold with the orifice tip. Tighten the nuts with your fingers.

The orifice hole size determines the diameter of the spray. In general, you should use a smaller orifice for smaller projects or for more detail and control. The length and consistency of your fiber will also determine the opening. A highly beaten short fiber will flow more easily through the smaller orifice hole than a longer, coarse fiber.

2. Make sure the sprayer valve is closed. Then connect the pulp sprayer to the compressor with an air hose. If your piece is large, use a long hose so that you can reach all of your work. In addition, if the compressor is noisy, you might want to use a long hose so that you can put the compressor in another room and still reach your piece.

3. Plug in the compressor. Set the compressor pressure at 60 to 70 PSI.

4. Make sure the trigger of the sprayer is in the "off" position. Then fill the hopper with as much beaten pulp as you can handle without overfilling or spilling (remember, wet pulp is heavy).

5. Aim the pulp sprayer away from your work, at something that is stable and won't blow away. Open the pulp sprayer air valve to the desired air flow volume. You'll have to judge this by experience — pay attention to the sound produced by the air flow. If the flow is too low, the pulp will come out of the sprayer intermittently or weakly, and you will get neither the volume nor the evenness you desire. If the flow is too high, it will blow the pulp off the surface you are spraying. Remember that the air flows through the center of the orifice continually when the air valve is open, so be careful not to aim the gun at your work — even with the trigger off, it can displace pulp you have already sprayed.

6. As you pull back on the trigger the air orifice retracts from inside the orifice plate hole. The pulp comes out between the orifice and the perimeter of the hole. The farther you retract the air orifice, the wider the gap and the more pulp released per second. The air flow remains constant. Therefore, the trigger determines the rate of pulp flow. Balancing the air flow (step 5) with the pulp flow (step 6) requires fine tuning and experience to get predictable results.

SPRAYING TECHNIQUES

Applying pulp to a horizontal surface is fairly simple. The easiest way to form an even layer of pulp is to spray at a right angle to the surface, or as close to perpendicular as possible. As you spray, use either a steady zigzag motion or a steady circular pattern, keeping the sprayer 3 to 6 feet from the surface of your piece so the air does not displace the pulp.

Varying Density

Spraying from an angle that is almost parallel to your piece causes an uneven buildup of pulp, which can be manipulated to create interesting textures. The pulp may form ridges that block the area behind

Spray at a right angle to a horizontal surface to form an even layer of pulp.

them, creating a wavelike variation of pulp density. These waves are clearly visible when the paper is backlit, but invisible when lit from the front.

You can control this technique by placing shallow objects under the fabric or screen material. Spray at an angle almost horizontal to the surface, and carefully remove the objects while the pulp is still wet (try attaching strings or thin wires to the objects so you can ease them out from beneath without displacing the pulp above).

Textural Variations

You can also fold, wrinkle, or crease the fabric or screen surface, spray the pulp, and then gently smooth the material out to its full area. This will pull the areas of pulp apart, leaving patches of wet pulp on the material. If you then spray another even layer on top, you can connect the pieces, creating a sheet with an interesting texture — especially visible when backlit. This method is best done with fine net material (nylon) or fabric, which can be laid on top of plastic sheeting anchored smoothly to the work table, or on top of a large piece of Plexiglas. If you have problems with drainage, tilt your surface slightly so that excess water can run off the edges.

Layering Pulp

You can combine different pulps in the same piece by layering. For example, you could spray one fiber or color first to create the inside of the piece, and then layer a second fiber or color on top to create the outside. If you layer different fibers on the same piece, pay particular attention to shrinkage.

In most cases, you will get the best results if you start by spraying a thick layer of a low-shrinkage pulp and then spray a thin layer of high-shrinkage pulp over it. The high-shrinkage pulp will adhere to the underlying layers and will thin and compress as it shrinks, forming molecular bonds both between the layers and within each layer. If the high-shrinkage pulp is the thicker of the two, it will pull significantly as it dries and may cause the low-shrinkage layers beneath to shrivel — particularly if the core or armature upon which it is applied has some give to it (like a nylon or foam armature).

Inclusions

You can incorporate inclusions such as flowers, seeds, threads, or ribbons between the layers of pulp as you spray, arranging them as you like. Dried organic matter works better than fresh, since living flower petals and leaves have water-repellent qualities. Apply the inclusions close to the surface layer if you want them to be visible when the piece dries.

Spray with care when applying the layer on top of your inclusions — the power of the compressed air can easily displace a petal or a feather. Gently tap them into the layer beneath them before spraying on top of them. In some cases, inclusions can deter pulp slides — they act like the terracing of farm fields on hillsides.

Encasing a Flat Object in Pulp

You can encase a piece of fabric, such as a piece of lace, by spraying it on both sides. To do this, stitch or anchor the fabric to a frame, such as a canvas stretcher bar frame, to suspend it between the sides of the frame. Consider shrinkage when doing this — use a strong thread or wire, and evenly distribute the tension around the frame.

To spray, set the mould on supports (such as two sawhorses) with one surface facing up. Apply a light, even coat to the top surface. If you plan to leave the frame attached to the paper, you might want to spray it with pulp as well so that it matches. Let the water drain through the fabric for a few minutes until it stops dripping. At this point it should look similar to a sheet of paper that has drained properly on a mould

To spray fabric on both sides, encase it in a frame.

and is ready to couch. Carefully flip the frame and apply a light, even coat to the other side. Take care to keep the sprayer far enough away, and spray a light enough coat so that you don't blow the pulp off the opposite side. Allow the second side to drain.

Flip the piece again, and apply a second coat of pulp to the first side of your piece. Repeat spraying layers until you reach the desired thickness. Two or three thin coats should suffice; too thick a coat will obscure the details of the lace. The fabric will be encased in pulp, and it will give the "sheet" of paper some strength. If you want to remove your piece from the frame, simply snip the threads or wires when it is dry.

Applying Pulp to a Vertical Surface or Form

This process is more tedious than working on a horizontal form. Gravity is your enemy, since applying a layer of pulp that is just slightly too thick can cause the pulp to slide. Draining water can also cause small streams that can wash pulp away. When working vertically, you must be patient and allow layers to evaporate just enough to allow the next layer to adhere (and the molecular bonds still form), yet dry enough to keep each subsequent layer from washing away the previous one. It is a challenge, but not impossible.

Large, vertical forms are especially challenging to keep consistent in thickness, since the top will inevitably dry faster than the bottom and will appear to need a new layer considerably sooner than the bottom. You can mist the top layer lightly with water (not enough to cause drips) to prevent it from drying, but you must do so before the surface has started to seal. If you are making a three-dimensional form and the areas do dry, subsequent layers can still be applied. However, you will end up with a piece made from several individual layers, and removal from the armature may be difficult. The inner layers may separate from the others and adhere to the armature itself.

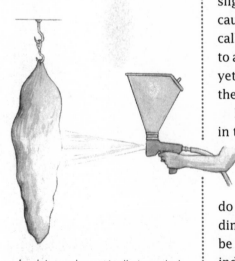

Applying pulp vertically is a challenge but can be accomplished with some patience and care.

Three-Dimensional Forms

Spraying on a three-dimensional hollow form (made of screen, for instance) requires care that the water and pulp sprayed through the form (after depositing pulp on the outside of the form) does not displace the pulp on the opposite side of the form. Prevent this by spraying light layers from an appropriate distance and allowing each layer to drain before you spray another layer. Remember that if pulp does spray through to the opposite side of the form, some pulp will be deposited on the inside of the opposite side of the armature — and if it bonds to the outside layer, removal of the armature will be impossible.

The underside of a three-dimensional piece is the most difficult place to spray — it is wetter than the rest of the piece, and gravity works against you. You can initially have the bottom side up, spray a few light layers (with draining intervals in between), then flip the piece and spray the rest of it. Be careful to prevent the edges of the first layers from drying out, or they will shrivel and shrink. The edges will initially be on the lower edge (when the piece is inverted), so gravity will aid in preventing premature drying, but it may be necessary to wrap them in strips of plastic to keep them from drying between layers.

LAYERING

Layering is necessary in almost all pulp-sprayed projects because you have to build up several thin layers to create a solid surface, except in two-dimensional objects sprayed directly onto horizontal moulds or onto plastic. When you form sheets of paper in a vat, you can vary the thickness by adding more or less pulp, but when spraying a vertical form, you can only apply so much pulp at a time. The time that elapses between applications will vary depending on your piece, the air temperature, and the humidity.

DRYING TECHNIQUES

To keep flat pieces from crinkling as they dry, clamp the fabric tightly to the table at the corners and, if the piece is large, at the midpoints on the edges. This technique works only for two-dimensional pieces.

Drying sprayed pieces, especially large ones, is best done slowly and naturally to avoid uneven drying, which can cause puckering and tearing. Tearing occurs especially if a wet area is adjacent to a dry one: As the shrinkage takes place in the final stages of drying, the wet pulp is very weak, and the dry area will simply pull the wet one apart.

You can use a fan, but I don't recommend blowing air directly on your piece. It is best to keep air circulating by deflecting it off a wall. Otherwise, you risk drying out one side of the piece prematurely. For large vertical pieces, keep a fan moving the air near the bottom of the piece, since the bottom always dries more slowly, and heat rises. If you can, rig up a fan to push ceiling air into an area of the perimeter of the room (not directly over the piece) to help equalize the temperature and circulate the air below.

Open Armature for Pulp Spraying

You can create a variety of sculptural forms that can be sprayed with pulp. An open armature is constructed so that it can be removed once the sprayed piece has dried, although you can choose to leave it attached. If you plan to remove the armature, you need to consider how, as well as think about pulp shrinkage and drainage. The following project uses wood and wire to create a ribbed structure from which you can cast a form with ridges similar to those found in a scallop shell or an acorn squash.

Clamping fabric as it dries keeps flat pieces from shrinking.

You Will Need:

- 1" thick sheet of plywood cut slightly larger than your form
- 9-gauge galvanized steel wire
- Drill and bits (one just larger than the diameter of your wire, and one just big enough for a needle to fit through)
- Wire cutters
- Brush for applying sealant
- Wood sealant, such as polyurethane
- Nylon mesh
- Fine, stretch netting with a knit weave (such as material designed for bridal veils, a lightweight cotton-knit fabric, or pantyhose)
- Tape
- Nonstretch nylon thread
- Sewing needle
- Prepared pulp
- Pulp sprayer and air compressor
- Scissors

TIP

Use a heavy-gauge, galvanized steel fencing wire, and check the wire's flexibility before using it. It should be stiff, yet flexible enough to work with. Shrinking paper can easily bend ¼" steel if the steel is too soft or is already bent in a shape that shrinking would bend further. A convex shape will resist bending whereas an "M" shape will collapse in upon itself.

1. Drill holes ½" to ¾" deep in the sheet of plywood, being careful not to drill all the way through the plywood. Drill them at an angle so that the lengths of wire will easily slip into them and not have to be bent upon leaving the holes, and make them a size into which the wires will fit snugly. Do not drill all the way through the plywood.

2. Cut lengths of wire to span the width of your piece, allowing extra length to fit into the holes in the plywood.

3. Place the wire ribs in the holes, bending them to create your desired form. Do not bend the wires just above the holes, because shrinkage may cause them to bend further and collapse. (Long, straight wires that run parallel to the plywood tend to do this.) Simple arches or wires bent at right angles tend to work best.

step 1

You can also weld support rods underneath, perpendicular to the arched wires and extending into the board, or you can weld pieces between the wires as a support structure. If you do this, however, these wires will make an impression in the paper as it shrinks. If your support wires are concavely bent, they won't show; but they will bend further and not do their job unless they are really strong.

4. Remove the rib wires (number them for easy replacement), and cut a slot or series of large holes in one end of the plywood for drainage, approximately 5" in from the edge of the plywood.

5. Drill small holes (to fit a needle through) at 1" intervals around the perimeter of the final piece; these will be used for sewing the mesh to the armature. To prevent warping, waterproof the board with wood sealant, including the insides of the drilled holes. When the board is dry, place the wires back in the plywood to create the armature.

6. Replace the wires, and sew the mesh onto the armature to create a surface. Lay the netting over the mould, fold it around all four edges to the back of the plywood, and tape it in place.

The tightness of the fabric depends on the desired final effect. If the fabric droops between the ribs, after being sprayed with pulp it will shrink and not droop as much when dry, because the shrinkage will flatten it out, tightening it between the ribs. If it is stretched tight, you will get more of a droop, especially if the fiber shrinks, because the paper will pull down between the ribs since it is less restrained. Keep in mind that highly beaten pulps opacify when allowed to shrink without restraint, so a tightly stretched fabric will yield the most translucent form.

7. Using nonstretch nylon thread and a sewing needle, stitch the netting to the plywood through the small drilled holes, using a long running stitch and keeping the thread tight. Use a strong thread that won't break as the paper shrinks. Catch the folded-back edges of nylon mesh underneath as you stitch, to anchor them in place.

step 6

8. Set the armature on a sawhorse or table in your prepared spray area so that you can walk around it completely. Raise the board slightly on the nonslotted end to facilitate drainage (with a decline of 1" to 2"). Spray the armature in layers. Be sure to plan the thickness of the piece carefully to ensure that it is thick enough to maintain its shape when finished. If it is too thin, the form will collapse and distort easily. Allow the piece to dry.

9. When the piece is dry, remove the armature by snipping the threads that anchor the fabric to the armature from beneath the board. Turn the piece over, and peel the fabric away from the paper (do not peel the paper from the fabric — you risk creasing the paper).

Variations

◆ The mould could be made with plywood ribs screwed to the base sheet of plywood, but the drainage of each section must be considered.

◆ You can make pieces using this method by crossing and looping wires to make your form. If you weld the points of intersecting wires, they will maintain their shape. If you wire or tie them together, there will be movement with shrinkage, and the form will flatten out. Your piece will also become more opaque because of shrinkage — as area decreases, density increases.

TIP

Artist Joyce Schutter recommends spraying primarily on the outside, convex surface of a mould, rather than the inside, concave side because the resistance to shrinkage is greater, which increases the translucency of the paper.

NYLON AND SAND ARMATURE

You can create another type of open armature by filling a nylon stocking with sand, suspending it from the ceiling, spraying the form with pulp, and then removing the sand and stocking after the paper dries. Use this technique to create interesting forms or component parts for assembled pieces.

You Will Need:

Nylon (pantyhose) or cotton-blend stretch stocking
Fine particulate material such as sand
Hook for suspension
Prepared pulp
Pulp sprayer and air compressor

1. Create a long, funnel-like form by filling a stocking with sand (you might need two or three layers of stockings to keep the sand from spilling through the holes).

2. Tie a knot in the open end of the stocking, and suspend the form from a hook.

3. Spray the piece with pulp in layers. The sand will absorb some of the moisture from the pulp, so make sure that both the paper and sand are dry before you drain the sand and remove the nylon.

4. Make a hole or untie the knot in the stocking, and drain the sand. This technique will work fine if the paper is fairly thick. If you have a thin layer, make a small hole in the bottom of your piece so that gravity drains the sand. You will still need an opening in the top of the piece to avoid creating a vacuum, which would collapse the paper.

Take care not to bend the sand-filled nylon armature with the dry paper on it, because the paper is likely to split.

step 2

MAKING OTHER SCULPTURAL FORMS

There are many other possibilities for creating simple to complex sculptural forms. Here are a few more ideas:

Balloons. You can spray onto the surface of a balloon, but the process is slow — very thin coats of pulp must be carefully applied, to prevent pulp slides. Build up enough layers of pulp to keep the paper from collapsing when the balloon is popped and removed.

Found Objects. You can spray onto any object that doesn't have holes or deep undercuts, which inhibit paper removal — see Avoiding Undercuts on page 163). This requires careful layering. Working over convex forms is best, so that the paper doesn't shrink away from the form as it dries. When the piece is dry, cut it along natural shifts in plane to make the cuts less visible, and reattach them.

Porous materials such as wood are easier to spray because the pulp will readily adhere to them, but paper removal will be more difficult. Nonporous materials such as plastic and metal work fine, although very thin layers of pulp will be difficult to remove. If you spray enough layers, your piece should be strong enough to maintain its integrity when pulled away from the underlying object — it will tend to stick to itself more than the object. Try various objects — you can even spray an old shoe!

Armatures. You can make forms of any size and shape using screen, wire, or wood and fabric (such as netting or cheesecloth). Just make sure that the armature is strong enough to bear the weight of the wet pulp and the power of the pulp's shrinkage. Sometimes just having suspension wires that hold the piece to hooks on the ceiling, the wall, or a frame will suffice.

CHAPTER 15

CHILDREN'S PROJECTS

Papermaking is a great educational activity for children. There are scientific properties to learn about (such as hydrogen bonding), a long history to discover (papermaking began in China in A.D. 105), and a variety of creative projects to undertake. Check out Gloria Zmolek Smith's book *Teaching Hand Papermaking: A Classroom Guide,* a thorough guide that includes instructions on how to set up a classroom for papermaking, a step-by-step papermaking guide to several techniques, and an entire section on relating papermaking to other school curricula.

Many of the techniques described in this book work well with children (ages 8 to 10 seem particularly fond of making paper). Making decorative papers, making shaped sheets, and laminating and embedding, are just a few projects that work well with children. Here are some tips I've found helpful when working with children, along with a few specific projects.

PAPERMAKING IN THE CLASSROOM

It can be a challenge to teach papermaking to large groups of children, but it's not impossible. If you have the luxury, I recommend working with small groups of four or five. You can have only so many vats in a room, and since the children cannot all make paper at the same time, you will have less crowding at the vat with fewer children.

If you do have a large group, ask extra teachers and parents to help out. Often, these adults are as excited as the children (if not more).

Space Set-up

Every classroom and group of students is different, so your papermaking setup will vary. It is ideal to have a sink in the classroom or at least nearby. Set up a limited number of vats and have children come up in small groups: one child forms a sheet, another lays a felt down, another waits to make the next sheet, and one helps the papermaker carry the sheet to the next station for decorating or pressing.

Having two or three vats on a table in the center of the room seems to work well. It is helpful to have one adult who is familiar with the process at each vat (they can come in a few minutes early for a briefing). Cover the tables with newspaper, and with plastic sheeting underneath if they need extra protection.

EQUIPMENT

The basic equipment described in chapter 2 is all suitable for making paper with children, but I'd like to mention a few additional things. First, work small. I recommend making sheets of paper that are 5½" x 8½" — an easy size for any age to master. Working small also means that all your other equipment (felts, presses, and drying systems) will take up less space, too.

BEATERS

A blender is a great tool for teaching, since kids can blend pulp themselves and watch it change from recycled paper, plant material, or preprocessed sheet fiber into pulp. Each child can tear up scraps of recycled paper and soak it in water before blending. A variety of other tools can be used to beat pulp (see chapter 3), but a blender is both sufficient and inexpensive.

BREAKING DOWN THE PAPERMAKING TASKS

There are several steps in the papermaking process, so you can put different groups of children in charge of the various stages, rotating as each group takes a turn at making paper.

◆ Setting up (filling vats, arranging stations)
◆ Tearing and blending various fibers
◆ Making name tags to identify sheets
◆ Assisting with papermaking (laying the felt for couching, helping the papermaker carry the sheet to the next station)
◆ Pressing
◆ Unloading the press
◆ Setting sheets of paper out to dry
◆ Doing water patrol and cleanup

TIP

It can come in handy to put a "this side up" label on the mould so children know which way to hold the mould. It is common for beginners to form the sheet on the wrong side of the mould, which makes couching impossible.

Hand-beating is educational and also uses up kids' energy. You will probably need to precook fiber that you will hand-beat, since cooking takes a couple of hours. Make sure the fiber is well cooked so the kids won't have trouble beating it. You will also need several mallets (I've used 1 x 2s with taped handles to prevent splinters and blisters), so that each child can pound a handful of pulp. Be aware that a classroom full of pounding is noisy, so you will need to either check into what is going on next door and below, if you are upstairs, or work outside. Mixing the plant fiber with another pulp is a good idea too, since the beating quality may vary.

SIMPLE MOULD AND DECKLE ALTERNATIVES

You can purchase simple papermaking kits at many art supply stores these days, and they are also available through papermaking and school suppliers. Alternative moulds and deckles include the following:

Embroidery hoop moulds: Simply stretch plastic window screening across an embroidery hoop, and make a mould for circular paper. You can make a deckle out of a second hoop, or use no deckle at all.

Deckle-less aluminum or needlepoint screen mould: Cut a piece of aluminum or needlepoint screening to the desired size. Use a waterproof tape to seam all four edges. This will protect hands from being scratched and provide a grip for making sheets. Try cutting the screen into shapes to make shaped papers, too. If your screen is too flexible, back it with a piece of ¼" hardware cloth cut to the same shape, and tape the two screens together.

Stretcher bar moulds: See the instructions for building one on page 14. Needlepoint bars will work, too.

Simple deckle box mould: See page 24.

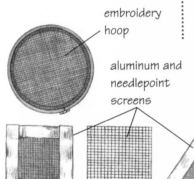

embroidery hoop

aluminum and needlepoint screens

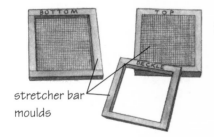

stretcher bar moulds

VATS

Many types of plastic bins make great vats. Consider purchasing containers with lids, which can double as storage containers for your papermaking supplies. When purchasing vats, just make sure that they are large enough to dip the mould and deckle into — you should have 3 inches to 4 inches all around for ease of movement, and the vat should also be at least 6 inches deep.

COUCHING MATERIAL

Again, any of the materials described in chapter 5 will work, but you will need to consider space and drying. Nonwoven interfacing is lightweight, dries quickly, and can be stored in a small space. You can even couch onto newspaper, which can be discarded once the sheets of paper are removed. If you are working in a school setting, cafeteria trays are great for couching because they collect excess water, which the students can then dump back into the vat.

Drying Techniques

You can air-dry sheets on a clothesline (make sure you sponge-press the sheets onto their drying material so that they are well attached before hanging) or in a storage shelf or rack. Each sheet will dry best if air is allowed to circulate all around it. Just remember that it could take a couple of days for the sheets to dry, so you will need sufficient space. If you have windows in the classroom, try brushing the sheets onto them, and enjoy the view as the sheets dry. Any of the drying methods described in chapter 5 will work also.

Brushing sheets onto the classroom windows lets you enjoy the view as they dry.

NONTOXIC COLORS

I have had success using tempera paints to pigment pulp. I use them when I'm working with children because they are nontoxic. Usually I start with a container full of beaten pulp, add a few squirts of tempera, and mix it into the pulp. I continue adding paint until the pulp reaches the desired color. Then I check the water for clarity, and if it isn't clear I add a few drops of retention aid and mix it again until the water is clear. (For more information on coloring, see page 77).

RECYCLING OLD PAPER

Perhaps the simplest project to do with a group is just making sheets of paper from recycled materials that students collect (old notebook papers, envelopes, letters, cereal boxes, tissue paper). Have students save junk mail or old school papers for a few days. They will then see that they are turning it back into a useful product.

You Will Need:

Recycled paper (you will need two or three sheets of recycled paper for every sheet you wish to make)
Bucket
Blender
Papermaking equipment

1. Cut or tear the recycled paper into 2" squares, and soak them in a bucket overnight. (You can also soak the paper and then tear it into squares.)

2. Blend the paper. This should be supervised by an adult, because many kids have never used a blender. And, since you are working with electricity and water, make sure the plug is safe from water exposure. Fill the blender three-quarters full, and add a handful of soaked paper. Turn on the blender, and run it until the paper has disintegrated. This is your pulp.

Caution: Too much paper with too little water can burn out your blender. Make sure the paper can move freely in the blender, and if your blender makes unusual noises, turn it off and unplug it. Remove some of the paper, add some water, and start again.

3. Dump the pulp into the vat, and continue blending batches until the vat is full. You may need to add extra water to the vat if the pulp is really thick. If it looks thin, strain some water out of the pulp after blending and before adding it to the vat.

4. Make paper!

WHAT TO RECYCLE

Some things to think about when collecting paper:

✦ Printing on paper will affect the color; if you want to make white paper, choose envelopes or office paper with little printing. If you want to make colored papers without coloring the pulp, add some colored tissue paper or construction paper.

✦ Glossy papers are more difficult to break down than unglossy.

✦ Remove any nonpaper parts, such as staples and plastic windows, from envelopes.

PAPER QUILT

One project described by Gloria Zmolek Smith, author of *Teaching Hand Papermaking* (see reading list), is a paper quilt. It's a nice project for a group of students because each student makes an individual square, and all the squares are joined to form a classroom quilt.

The quilt can be displayed at the school and later disassembled so each student can take home his or her individual square to show off to the family.

You Will Need:

Prepared pulp (any pulp will work)
Papermaking equipment
4 strings per student

Making Individual Quilt Squares

Set up the classroom for papermaking. Each student follows these steps when it is her/his turn to make paper.

1. Make a sheet of paper, and couch it on a couching surface.

2. Double-couching is the best way to securely embed strings in the quilt squares (see box on page 202). Have the student place strings on top of the freshly couched sheet, as shown in the diagram. Students should center the strings so they are long enough on each end to be tied together when they dry. Next, the student should make a second sheet of paper and couch it directly on top of the first sheet.

3. After creating the quilt square with strings, have the students decorate their squares with pulp painting, collage, string paintings, or any other technique described in the previous chapters.

4. Press and dry the sheets.

5. Arrange the panels, and tie them together with the embedded strings, creating the quilt.

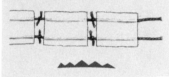

VARIATION

To make a large accordion book, place two horizontal strings on each square so the sheets can be tied together in a row.

step 2

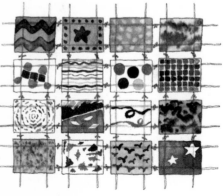

step 5

DOUBLE-COUCHING AND ONE ALTERNATIVE

When couching one sheet of paper on top of another, you will need a guide or jig for lining up the sheets. Before making any paper, place the mould face down on the couching surface (a). Mark two adjacent corners with waterproof tape so that the mould is placed in the same spot each time a child couches.

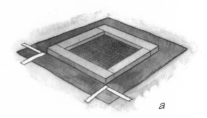
a

Double-couching can be tricky, and it takes more time for each student to form two sheets. An alternative is to have students dip the strings one at a time into some watery pulp (b) and then lay the strings onto the wet sheet of paper. Any fibers that attach to the string will bond with the fibers on the wet sheet. There is

b

no need to form a second sheet of paper when this method is used. Try dipping the strings in colored pulp for an added dimension. Students can also spoon or squirt watery pulp over the string to embed it.

PAPERMAKER'S PROJECT

BOOKMAKING

Here is a simple book project you might try with your students. Have them use the books as journals or notebooks — they can even use them to document the papermaking process.

You Will Need:

8½" x 11" handmade paper covers (flat size)

5 sheets of 8½" x 11" text paper per student (handmade or office paper)

Bone folder, pen, or plastic knife

Embroidery needle or awl

Embroidery floss

Preparing the Cover

1. Have students create book covers using any technique, such as collage or pulp painting. You can make them any size you like, but I like to make the unfolded size of my covers either 8½" x 11" or half that size: 5½" x 8½", so that the inside pages can be made out of regular-size 8½" x 11" paper.

When the students decorate their sheets, have them think about the front and back covers of a book, because they will be making a flat sheet that folds in half to become a book cover. You can have them sketch their ideas beforehand, or even while they are waiting for a turn to make paper.

2. If you like, you can have them also make the text sheets for the book in plain white paper. Keep the text sheets the same size as the cover, or slightly smaller. Otherwise, take standard 8½" x 11" paper, or cut it in half to measure 5½" x 8½".

3. Have each student take five sheets of text paper and fold them in half. Score (crease) the signature (the group of folded pages) with a bone folder, the back of a pen, or a plastic knife.

step 1

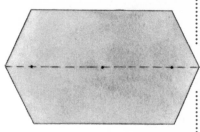

4. I make punching cards ahead of time for punching holes for sewing. The length of the card needs to equal the height of the signature (8½" or 5½"). You can also have students make them themselves. To punch, fold the punching card in half with the markings on the inside. Have each student prepunch the punching card with an embroidery needle or an awl. Then place the card in the center of the folded signature so that the ends line up, and punch through the entire signature. Tell students to be careful because the awl end is sharp. Repeat this process to punch holes in the cover.

5. Have students thread the embroidery needles with embroidery floss, which is inexpensive and comes in many colors. I usually cut it into lengths before giving it to the students, or have a predetermined length for them to cut. Many students have never threaded a needle, so you may need extra time to help them.

Stitching the Books

1. Have students place the cover around the text and then open up the text and cover. Starting from the outside, stitch into the cover and text through the center hole, leaving a tail 3" long.

2. On the inside, sew through and out one of the other holes.

3. On the outside, bring the thread all the way to the hole that hasn't been stitched into, and sew into it.

4. Bring the thread back out through the center hole. When doing this, note that the tail is sitting on one side of the length of thread that spans the spine. Bring the new end of thread to the opposite side of that length. Remove the needle, and tie a knot in the two ends of thread.

5. Snip the ends of thread so that they are approximately ½" long.

CONVERSION CHART

All measurements in this book are given in the English system. But the metric system, used internationally, is handy for weighing out ingredients. This simple conversion chart will help you convert between the English and metric systems when you need to.

To convert to	When the measurement given is	Multiply it by
milliliters	teaspoons	4.93
milliliters	tablespoons	14.79
milliliters	fluid ounces	29.57
milliliters	cups	236.59
liters	cups	0.236
milliliters	pints	473.18
liters	pints	0.473
milliliters	quarts	946.36
liters	quarts	0.946
liters	gallons	3.785
grams	ounces	28.35
kilograms	pounds	0.454
centimeters	inches	2.54
degrees Celsius	degrees Fahrenheit	$\frac{5}{9}$, then subtract 32

RECOMMENDED READING

Adrosko, Rita J. *Natural Dye and Home Dyeing.* Magnolia, MA: Peter Smith Publisher, Inc., 1971.

Barrett, Timothy. *Japanese Papermaking* (with an appendix on alternative fibers by Winifred Lutz). New York: Weatherhill, 1983.

Bell, Lilian A. *Plant Fibers for Papermaking.* McMinnville, OR: Liliaceae Press, 1981.

———. *Papyrus, Tapa, Amate and Rice Paper: Papermaking in Africa, the Pacific, Latin America and Southeast Asia.* McMinnville, OR: Liliaceae Press, 1983.

Britton, Nathaniel Lord, and Hon. Addison Brown. *An Illustrated Flora of the Northern United States and Canada* (in three volumes). New York: Dover Publications, Inc., 1970.

Casselman, Karen Leigh. *Craft of the Dyer: Colour From Plants and Lichens.* New York: Dover Publications, Inc., 1993.

Crawford, William. *Keepers of Light: A History and Working Guide to Early Photographic Processes.* Dobbs Ferry, NY: Morgan and Morgan, Inc., 1979.

Dawson, Sophie. *A Hand Papermaker's Sourcebook.* London: Estamp, 1995.

———. *The Art and Craft of Papermaking,* Philadelphia: Running Press, 1992.

Farnsworth, Donald. *A Guide to Japanese Papermaking: Making Japanese Paper in the Western World.* Oakland, CA: Magnolia Editions.

Grummer, Arnold. *Paper by Kids.* Minneapolis: Dillon Press, 1989.

———. *Tin Can Papermaking: Recycle for Earth & Art.* Milwaukee, WI: Greg Markim Publishers, 1992.

Heller, Jules. *Papermaking.* New York: Watson-Guptill, 1978.

Hughes, Sukey. *Wash: The World of Japanese Paper.* Tokyo: Kodansha International, 1978.

Hunter, Dard. *Reprint of Papermaking in the Classroom.* New Castle, DE: Oak Knoll Press, 1931, 1991.

———. *Papermaking: The History and Technique of an Ancient Craft,* 2nd ed. New York: Dover Publications, Inc., 1978.

LaPlantz, Shereen. *Cover to Cover: Creative Techniques for Making Beautiful Books, Journals & Albums.* Asheville, NC: Lark Books, 1995.

Maiti, R. K. *Plant Fibres.* Dehra Dun: Bishen Singh Mahendra Pal Singh, 1980.

Mason, John. *Paper Making as an Artistic Craft,* revised edition. Leicester: Twelve by Eight Press, 1963.

McRae, Bobbi A. *Colors From Nature: Growing, Collecting & Using Natural Dyes.* Pownal, VT: Storey Communications, Inc., 1993.

Miller, Dorothy. *Indigo from Seed to Dye.* Aptos, CA: Indigo Press, 1981.

Premchand, Neeta. *Off the Deckle Edge.* India: The Ankur Project, 1995.

Robbert, Paul. *"The Construction of Vacuum Tables: Theory and Practice."* Hand Papermaking, Vol. 6, No. 1, summer 1991.

Rudin, Bo. *Making Paper: A Look into the History of an Ancient Craft.* Väallingby, Sweden: Rudins, 1990.

Sandberg, Gosta. *Indigo Textiles Technique and History.* Asheville, NC: Lark Books, 1990.

Shannon, Faith. *The Art & Craft of Paper* (formerly *Paper Pleasures*). New York: Weidenfeld & Nicolson, 1994.

Smith, Gloria Zmolek. *Teaching Hand Papermaking: A Classroom Guide.* Cedar Rapids, IA: Zpaperpress, 1995.

Stearns, Lynn. *Papermaking for Basketry.* Asheville, NC: Lark Books, 1992.

Sward, Marilyn, and Catherine Reeves. *The New Photography.* New York: Da Capo Press, 1996.

Toale, Bernard. *The Art of Papermaking.* Worcester, MA: Davis Publications, Inc., 1983.

Turner, Silvie. *Book of Fine Paper.* New York: Thames and Hudson, 1998.

Turner, Silvie, and Birgit Skiöld. *Handmade Paper Today: A Worldwide Survey of Mills, Papers, Techniques and Uses.* London: Lund Humphries Publishers Ltd., 1983.

Watson, David. *Creative Handmade Paper: How to Make Paper from Recycled and Natural Materials.* Kent, England: Search Press, Ltd., 1991.

The following books are self-published or limited editions that might be difficult to find, but you may track them down through papermaking suppliers, in special collections, or via the World Wide Web. If you are really interested in learning more about this topic, these books are worth the search.

Alexander, Harold H., and Marjorie A. Alexander. *Handcrafted Paper & Paper Products Made from Indigenous Plant Fibers.* Arden Hills, MN: Maralex Studios, 1997.

Koretsky, Elaine. *Color for the Hand Papermaker.* Brookline, MA: Carriage House Press, 1983.

Koretsky, Elaine, and Donna Koretsky. *The Goldbeaters of Mandalay.* Brookline, MA: Carriage House Press, 1991.

Jager, Edwin. *How to Make Big Sheets.* Toronto/Iowa City: My Name is Edwin Press, 1996.

McDonald, Lee, ed. *Beater Builders of North America 1946–1989.* Chillicothe, OH: The Friends of Dard Hunter, 1990.

Reeves, Dianne L. *From Fiber to Paper.* Austin, TX: Dianne L. Reeves, 1991.

Richardson, Maureen. *Vegetable Papyrus.* England: Berrington Press.

Siegenthaler, Fred. *Strange Papers, A Collection of the World's Rarest Handmade Papers.* Muttenz, Switzerland: Fred Siegenthaler, 1987.

RESOURCE GUIDE

Mail-Order Suppliers

Canoy Studios
8570 NW Marshall,
Portland, OR 97229
(503) 292-6205; (888) 260-3889,
  pin 1864
fax: (503) 291-1151
wet pulps

Carriage House Paper
79 Guernsey Street
Brooklyn, NY 11222
(718) 599-PULP; (800) 669-8781
fax: (718) 599-7857
papermaking supplies, equipment, custom paper

Dieu Donné Papermill, Inc.
433 Broome Street
New York, NY 10013
(212) 226-0573
fax: (212) 226-6088
e-mail: ddpaper@cybernex.net
Website:
www.colophon.com/dieudonne/
custom papers, papermaking supplies, classes and workshops, gallery

Guerra Paint & Pigment
510 E 13th Street
New York, NY 10009
(212) 529-0628
fax: (212) 529-0787
pigments

Richard Hungerford
P.O. Box 113
Keswick, IA 50136
(319) 339-2206
Website: www.netins.net/show
  case/artpaper

The Lamp Shop
P.O. Box 3606
Concord, NH 03302-3606
(603) 224-1603
fax: (603) 224-6677
Website: www.lampshop.com
lamp-making supplies

Julie Lau
e-mail: jaljgdp@spiritone.com
templates and instructions for various designs, including bonbon box

Lee Scott McDonald
P.O. Box 264
Charlestown, MA 02129
(617) 242-2505; (888) 627-2737
fax: (617) 242-8825
papermaking supplies, student and professional equipment, specialized production equipment

Gold's Artworks
2100 N. Pine
Lumberton, NC 28358
(800) 356-2306
phone/fax: (910) 739-9605
papermaking supplies

Green Heron Book Arts
1928 21st Avenue Suite A
Forest Grove, OR 97116
(503) 357-7263
e-mail: bookkits@aol.com
Website: www.green-heron
  kits.com
papermaking and bookbinding kits and supplies

A. Leventhal & Sons, Inc.
711 Davis Street
Scranton, PA 18505
(800) 982-4312
dental syringes for pulp painting

Magnolia Editions
2527 Magnolia Street
Oakland, CA 94607
(510) 839-5268
fax: (510) 893-8334
papermaking supplies, equipment, books

Timothy Moore Paper Molds and Bookbinding Tools
14450 Behling Road
Concord, MI 49237
(517) 524-6318
high-quality professional papermaking and bookbinding equipment

The Papertrail
170 University Avenue West
Waterloo, Ontario
Canada N2L 3E9
(800) 421-6826; (519) 884-7123
fax: 519-884-9655
papermaking supplies

H. H. Perkins Co.
10 South Bradley Road
Woodbridge, CT 06525
(800) 462-6660
fax: (203) 389-4011
Website: www.hhperkins.com
reeds and basketry supplies (for lamp project)

PRO Chemical & Dye, Inc.
P.O. Box 14
Somerset, MA 02726
(888) 2-BUY-DYE
fax: (508) 676-3980
Website: www.prochemical.com
dyes, craft supplies, books

David Reina Designs, Inc.
79 Guernsey Street
Brooklyn, NY 11222
(718) 599-1237
fax: (718) 599-7857
Hollander beaters, presses, and drying systems

Talas
568 Broadway, Suite 107
New York, NY 10012
(212) 219-0770
fax: (212) 219-0735
e-mail: info@talas-nyc.com
Website: www.talas-nyc.com
bookbinding supplies

Twinrocker
P.O. Box 413
Brookston, IN 47923
(765) 563-3119; (765) 563-3210;
(800) 757-8946
fax: (765) 563-8946
e-mail:
    twinrock@twinrocker.com
Website: www.twinrocker.com
papermaking supplies

Watson Paper Company
1719 Fifth Street NW
Albuquerque, NM 87102
(505) 242-9351
fax: (505) 243-5644
handmade paper, paper products, papermaking supplies

Harry Wold
P.O. Box 5202
Aloha, OR 97006
(503) 641-2294
fax: (503) 641-7694
custom-built stampers

Jana Pullman
Western Slope
P.O. Box 7663
Minneapolis, MN 55407
(612) 722-9740
papermaking tools and equipment

Publications

Book Central
P.O. Box 895
Cairo, NY 12413
(518) 622-0113
e-mail: bookcentral@artist-
    books.com
Website: www.artistbooks.com
sells books on topics ranging from traditional bookbinding to contemporary book structures, papermaking, gocco printing, rubber stamping, pop-ups, and creating books with children

Hand Papermaking
P.O. Box 77027
Washington, DC 20013-7027
(800) 821-6604; (301) 220-2393
fax: (301) 220-2394
e-mail: handpapermaking@
    bookarts.com
Website: www.bookarts.com/
    handpapermaking
journal of contemporary papermaking, with a comprehensive listing of places to take classes, purchase supplies, and find out what is happening in the field

Somerset Studio
22992 Mill Creek Road, Suite B
Laguna Hills, CA 92653
(949) 380-7318; (877) 782-6737
fax: (949) 380-9355
Website: stampington.com
magazine featuring paper, letter-
ing, rubber stamping, and book
arts techniques and artists

The Book Arts Classified
P.O. Box 77167
Washington, DC 20013-7167
(800) 821-6604
fax: (301) 220-2394
e-mail: pagetwo@bookarts.com
Website: www.bookarts.com
newsletter listing recent publica-
tion of hand-crafted and trade
publications, classes and work-
shops, meetings, lectures and
conferences, exhibitions, online
sites, and more. Includes paper-
making listings

Papermaking Organizations

Friends of Dard Hunter
c/o Dard Hunter III
P.O. Box 771
Chillicothe, OH 45601
Website: www.slis.ua.edu/ba/
dardo.html
national membership organiza-
tion comprised of members inter-
ested in all facets of papermak-
ing; publishes a newletter and
holds annual meetings

International Assocation of
Hand Papermakers and
Paper Artists
c/o Kitz Rickert, Co-treasurer
2443 West Sunnyside #1
Chicago, IL 60625
Website: www.design.dk/org/
iapma
international membership organi-
zation comprised of members
interested in all facets of paper-
making

Paper Road/Tibet
3724 McKinley St. NW
Washington, DC 20015-2510
(202) 966-4828
fax: (202) 244-5952
Website: www.paperroadtibet.org

Robert C. Williams American
Museum of Papermaking
500 10th Street, NW
Atlanta, Georgia 30318-5794
(404) 894-7840
fax: (404) 894-4778
e-mail. Cindy.Bowden@ipst.edu
Website: www.ipst.edu/amp/
papermaking museum housing
the information, materials, and
tools that Dard Hunter collected
during his lifetime

Historic Rittenhouse Town,
Inc.
206 Lincoln Drive
Philadelphia, PA 19144
(215) 438-5711 (main office)
(215) 843-2228 (papermaking
studio)
fax: (215) 849-6447
e-mail: HistRitTwn@aol.com
birthplace of papermaking
in America; provides tours and
papermaking workshops

The Research Institute of
Paper History
Carriage House
8 Evans Road
Brookline, MA 02146-2116
617-232-1636
fax: 617-277-7719
e-mail: paperroad@aol.com
nonprofit organization housing a
papermaking facility; a library of
old, modern, and rare books on
paper history; and equipment
and artifacts relating to historical
papermaking from around the
world

INDEX

Page numbers in *italic* refer to illustrations; those in **boldface** refer to tables.

OTHER STOREY TITLES YOU WILL ENJOY

Papermaking with Plants: Creative Recipes and Projects Using Herbs, Flowers, Grasses, and Leaves, by Helen Hiebert. With step-by-step instructions, readers learn how to collect and harvest plant fibers from stalks, bark, leaves, and grasses; process the fiber and press, dry, and finish paper using Eastern and Western methods; embellish paper with natural dyes and decorative materials such as flower petals and pine needles; and craft one-of-a-kind projects. 112 pages. Hardcover. ISBN 1-58017-087-0.

The Handmade Paper Book, by Angela Ramsay. Easy-to-follow recipes for completing over 20 exquisite paper projects. Create unique and stylish personal stationery; add texture and color using natural materials such as banana, silk, flowers, and feathers; produce textured gift wrap and laid fiber paper; and master specialty papermaking techniques including embossing and watermarks. Full-color photos throughout. 80 pages. Hardcover. ISBN 1-58017-174-5.

Making Your Own Paper: An Introduction to Creative Papermaking, by Marianne Saddington. Step-by-step instructions and color illustrations provide the beginner with information about using a mold, pressing and drying, coloring and texturing, preparing a writing surface, and creating paper arts and crafts. 96 pages. Paperback. ISBN 0-88266-784-X.

Nature Printing with Herbs, Fruits & Flowers, by Laura Donnelly Bethmann. Step-by-step instructions for applying paint directly to plants and flowers to press images onto stationery, journals, fabrics, walls, furniture, and more. 96 pages. Hardcover. ISBN 0-88266-929-X.

The Aromatherapy Companion, by Victoria H. Edwards. A comprehensive guide to using essential oils and carrier oils. Hundreds of soothing and healing recipes for beauty, health, and emotional well-being; includes blends for the bath, massage, perfumery, and aphrodisiacs, as well as special recipes inspired by mythology, astrology and Ayurvedic medicine. 288 pages. Paperback. ISBN 1-58017-150-8.

The Candlemaker's Companion: A Comprehensive Guide to Rolling, Pouring, Dipping, and Decorating Your Own Candles, by Betty Oppenheimer. Guides readers through both the how and why of candlemaking, with step-by-step instructions for creating rolled, poured, molded, and dipped candles as well as specialty techniques such as overdipping, painting, layering, and sculpting. 176 pages. Paperback. ISBN 0-88266-994-X.

The Soapmaker's Companion: A Comprehensive Guide with Recipes, Techniques, and Know-How, by Susan Miller Cavitch. An authoritative handbook for making natural vegetable-based soaps, emphasizing specialty techniques for making marbled, layered, transparent, liquid, and imprinted bars. Covers chemistry, ingredients, additives, colorants, and scents, and explains how to experiment with fats and oils. 288 pages. Paperback. ISBN 0-88266-965-6.

These books and other Storey Books are available at your bookstore, farm store, garden center, or directly from Storey Books, Schoolhouse Road, Pownal, Vermont 05261, or by calling 1-800-441-5700. Or visit our Website at www.storeybooks.com